YOU AND THE NEW NORMAL

DR MOHAMED BUHEJI &
FUTURIST CHET W. SISK

authorHOUSE

AuthorHouse™ UK
1663 Liberty Drive
Bloomington, IN 47403 USA
www.authorhouse.co.uk
Phone: 0800 047 8203 (Domestic TFN)
*   +44 1908 723714 (International)*

© 2020 Dr Mohamed Buheji & Futurist Chet W. Sisk. All rights reserved.

No part of this book may be reproduced, stored in a retrieval system, or transmitted by any means without the written permission of the author.

Published by AuthorHouse 05/28/2020

ISBN: 978-1-7283-5339-5 (sc)
ISBN: 978-1-7283-5338-8 (e)

Print information available on the last page.

Any people depicted in stock imagery provided by Getty Images are models, and such images are being used for illustrative purposes only. Certain stock imagery © Getty Images.

This book is printed on acid-free paper.

Because of the dynamic nature of the Internet, any web addresses or links contained in this book may have changed since publication and may no longer be valid. The views expressed in this work are solely those of the author and do not necessarily reflect the views of the publisher, and the publisher hereby disclaims any responsibility for them.

CONTENTS

Preface vii
What is a 'new normal'? ix

Chapter 1   COVID 19. Opening the
            door to the new normal. 1
Chapter 2   Socioeconomic spillovers
            shaping the new normal...........................8
Chapter 3   Seven new ways of thinking
            in the new normal. 33
Chapter 4   How the new normal will
            deepen our relationships. 58
Chapter 5   The future of work in the new normal... 64
Chapter 6   Re-evaluating belief in the new normal...73
Chapter 7   We have some decisions to
            make about common practices................78
Chapter 8   Why do we seem to be
            terrible at change? 93
Chapter 9   The emerging role of
            community in the new normal. 97

Chapter 10	Preparing for the wave of challengers to the new normal.	117
Chapter 11	A scientific breakdown of how we can manage future disease and crises in the new normal.	122
Chapter 12	Re-thinking poverty in the new normal.	158
Chapter 13	Global risks in the new normal.	186
Chapter 14	Why curiosity is so important in the new normal?	210
Chapter 15	A philosophical take on this moment in time.	228
Chapter 16	What comes after surviving human complex challenges?	239

Brief about the Authors 243

Preface

Socioeconomist Dr. Mohamed Buheji
Futurist Chet W. Sisk

WHAT IS A 'NEW NORMAL'?

Around two million years ago, a young species of proto-humans emerged on the great plains of Africa. At the same time that these four-foot-tall, bipedal beings appeared, a period of climate crisis occurred. Instead of global warming, it was global cooling. It did not occur due to human activity, but to a black swan event that has yet to be fully understood. This climate crisis event ultimately forced these ancestors to adapt to a new way of life or perish entirely. They chose to change and live. According to the work of Michael Tomasello, a psychologist at the Max Plank Institute of Evolutionary Anthropology, the anthropology evidence gathered in a 2011 peer-reviewed study suggests major changes in the way these ancient ancestors operated. They moved from war-like individual clans to interdependent communities. They repelled members of the clan that acted individually instead of what was good for the group. They leaned on a cooperative community model to find, hunt and forage for food. In other words, they chose to do something about the coming crisis both for themselves and other clans. They elevated their behaviour to meet the moment. They embraced a 'new normal'.

We too are now in a new normal moment and are deciding if we want to embrace it, or revert back to the status quo. Of course, reverting back would be turning our back on an evolutionary urging for us to become better, higher, stronger. The current crises are giving us a signal. It's hard to find a time in history when so many elements of change were all happening at the same time --- a virus pandemic that impacts the world, climate change crisis that threatens the very existence of everything and everyone on earth, a change from the dominance of the West in world affairs to a more multilateral one, a widening income inequality gap that is creating a polarized world unlike anything we've ever seen, and the list goes on. There is no other time in recorded history that matches this unique moment.

The new normal is a disrupted space in time that requires a new set of rules and ideas in order to operate. For many of you, having your first child created a new normal. For others, retirement has. The COVID-19 pandemic abruptly introduced us to a new normal that has been decades in the making. The pre-covid world has been showing signs of malfunction for eons in the lives of millions. The recent pandemic merely brought the challenges of poverty, systemic racism, and the loss of jobs to automation, climate crisis and social breakdown to the fore. The new normal is both permanent and transient. It signals a new world has emerged and suggests another one is being born in this period, depending on what decisions we make at this time.

One such new normal in history many are familiar with was the 1789 French Revolution. Top down oligarchy control proved to be a bad management choice for the French ruling class. A new normal was ushered in as the leaders of this movement sought to figure out the single question "can we do better than this?" Back then, the French Revolution was primarily a regional event that affected certain aspects of the West. In today's world, a revolution of that magnitude would have global ramifications. This is what makes every world event now intimate. The world is quite connected. Even if there are those who don't believe this truth, ask them for a tour of their home. Check the labels on their things, and you'll see a global cornucopia of items: clothes from China, coffee from Ethiopia, electronics from Japan, cars from South Korea, beer from Mexico, art from Nigeria, pharmaceuticals from Germany. Trade, conferences, and even more and more dating successes have a global stamp on them. No other version of earth can make this claim.

This book recognizes how everything is connected and how the current nexus of major disruptions are changing the world that we once knew. We, as authors of this book, have used our skills, insight and experience to describe the particulars of this new normal and a blueprint of how we can leverage this moment to transition to a better and empowered planet. We can employ the connected nature of this new world, as well as the growing global hunger for new ideas, new innovations and new concepts of what can be. This is a time brimming with great opportunities for the greater good. But there are also challenges. This book seeks to carefully lay out both possibilities and pitfalls, but always through a "this is what we can do" lens. Success in this new world requires another kind of thinking…not just the thinking of scale, but of vision and a new way forward.

Our great blind spot, collectively as humans, is our constant desire to hold on to old ideas when the world around us is asking us, *pushing* us to change and adapt. This push is called evolution. In history, we regularly get to this next evolutionary place, but it's usually a bloody mess along the way. We may be able to avoid that trap this time by developing and implementing a new infrastructure for how things work -- first in our communities, then in our cities and perhaps our nations.

In the pre-covid old model, exploitation of the environment and people has led us to the current abyss.

That means whatever we did leading up to this moment needs to change. It's really that simple.

RE-EVALUATING THINGS THAT DON'T WORK.

There is an old African proverb that is most appropriate for this time. There was a village on the river. One day, one of the citizens saw several babies floating in the river. He jumped in the river to save them, and then saw several more coming down stream. He realized this was a trend. He then called the mayor, who suggested that they form a task force to start saving the babies from the river. A task force was formed to create a tool that saved the babies from the river. For months, the task force and the system they created to retrieve the babies from the water was the toast of the town. The task force received awards and more grant money so they could do more of this work. There were dinners, ceremonies and news coverage from all around. One day, a visiting woman saw the process of the task force saving the babies from the river, then asked the leader of the task force the most important question of that time: *who is throwing the babies in the river upstream creating the crisis in the first place?* The new normal isn't just about change for change sake. It's about doing the deep dive by asking questions like "why are we having these crises in the first place?" "Who or what is causing them?" "What can we do to end this?" "Is there a better way forward?"

"This book is about you asking and answering those questions while in the new normal of the COVID-19 affected world. We cover your relationships, your personal goals, your jobs, your beliefs and serious determinants at work in the world. For those who believe this is just another usual crisis to get through so we can return to what we left, may I submit exhibit A: we are facing an extinction level event called climate change caused by us. Whatever we've been doing personally is not enough and whatever we've been doing collectively is failing.

In essence, the coming new normal contains five major areas that are shifting in front of our very eyes. They are:

1. The leadership models that brought us here are not appropriate or agile enough for the next phase of our existence. In the words of Albert Einstein "we cannot solve our problems with the same thinking we used to create them."
2. Embracing the connected nature of the natural world and our new technology can help us create better systems in infrastructure development, effective leadership models, financial stewardship, partnerships and agreements.

3. We must rethink our beliefs that drive our current socioeconomic systems and embrace new adaptive ones that would work more efficiently for all people and the planet.
4. There are new opportunities emerging in this moment, but it will take visionary people with foresight to see them.
5. We must reorganize how our communities operate, empower them and network them into a nimble and flexible force for good in the world.

<u>You and the New Normal</u> will explain tools like future foresight, adaptive thinking and new economic models, while remembering that this is all about your relationships, your opportunities to make a living and the future of generations to come. To manage this time of mega global transformations, we'll need all shoulders at the wheel….more people participating in life-affirming, forward thinking actions that look out for all of us, especially our most vulnerable. We submit in this book, that a socioeconomic model that empowers more people to do well is more efficient, more effective and more abundant that models that don't.

As history has shown, change has never been our best subject in class. We haven't formalized the process so it can quickly devolve into a haphazard, helter-skelter affair. What we're suggesting, by the case studies under Socioeconomist Dr. Buheji and by the observational insight from Futurist Chet W. Sisk, are that we can do transformation better. When we do, we open up to opportunities that we never realized under the old way of thinking. My job is to constantly remind those I speak to that change isn't about loss, but rather about gaining that we've lost over the years in inefficiency, missteps and lack of imagination.

THE ESSENCE OF THIS BOOK

There are hidden and not-so-hidden opportunities in this new normal. We drill down on them so that you can use this book as a general guide in your management practices, organizational development and personal lives in a space that looks drastically different than what you knew just last year. This book joins a chorus of people, from the political arena to hedge fund managers to activists all singing the same song….'can we do better than this?' The obvious answer is, of course we can.

We have started our job of providing very compelling, well-researched and hopefully, inspiring blueprint for this new normal. You will have to summon the courage to do something with it.

CHAPTER 1

COVID 19. Opening the door to the new normal.

We are currently facing challenges many thought we would never see in our lifetime. These crises threaten our ability to move into the future. A perfect storm of events (climate crisis, pandemics, tech evolution and economic injustice, etc) have conspired to create, what we would call a socioeconomic bottleneck. Much like the crisis of 2 million years ago, we too are being asked to transform to meet the demands of a new world. To the average citizen of means, this demand seems far off…distant from the present world of trade, globalism and commerce. To the billions who are a small financial emergency away from economic ruin, this nameless, faceless, force is not only real, but it haunts our communities like an unrelenting ghost. This level of stress and precarious existence has helped us develop a new term --- the precariat --- a social group lacking the financial and psychological securities of existence. Even with the emergence of the precariat, our leaders seem not to care about this level of anxiety in our word.

Then, the COVID-19 pandemic of 2020 happened. This unseen but *predicted* event rocked the world's foundation by putting a stress test to our current systems. Unfortunately, many of our structures throughout the world failed, resulting in the deaths of thousands and took our communities to the abyss of economic collapse. This event is, again, asking us to elevate our behaviour in order to transform into something greater…..something that is robust and agile enough to get humanity through this bottleneck.

We are going to name several essential skills and tools in this book we must make standard in our communities, our businesses, our education institutions, our relationships and in our leadership models to meet the demands of a new world emerging. COVID-19 has made it clear that you cannot "pour new wine into old wine skins". Another world is seeking to emerge, but it needs an entirely new frame of reference. This is not to say we must throw out "the baby with the bathwater", but rather build upon the best practices of the past and expand on our capacity for success. This not only takes a new mind set but a new set of approaches, which we will outline in this book. These approaches, often called "soft skills" by the fading world, are the new determinants of success of the new normal and beyond.

The world is requiring another kind of leadership model. The often-used top-down pyramid "command and control" structure of leadership and its accompanying tools (control, elements of fear, aggressiveness, etc.) is simply not agile enough to meet the demands of a profoundly different future. This new model can be called the **Ubuntu Socioeconomic Construct,** (named after the African philosophy). It is a model focused more in keeping with a new and inspired future and is strong enough to face a world of new threats while offering new ways of approaching old challenges. Before we share these new approaches, we'd like to offer some context for the COVID-19 pandemic that has shaped the global narrative of 2020.

WHY IS COVID-19 DIFFERENT FROM OTHER PANDEMICS?

Make no mistake; the COVID-19 crisis is a deadly one. It has killed hundreds of thousands around the world and has brought illness to millions and has the potential of becoming as tragic as the Bubonic plague that led Europe into its dark ages. However, let's be sure to put this challenge into current perspective by presenting it alongside other pandemics.

TABLE (1) BRIEF HISTORY OF PANDEMICS

Crisis	Time Period	Global Death Toll
The Bubonic Plague	1346-1353	200 million
The Spanish Flu Pandemic	1918	50 million
The AIDS/HIV Pandemic	2005 - 2012	36 million
The Bubonic Plague	542 AD	25 million
The Antionne Plague	165 AD	5 million

As you can see, the COVID-19 pandemic, at the current trajectory isn't in the top 5 of global pandemics. In fact, as it is being measured right now, it may not make the top 10 pandemics of history when it comes to the death toll. So why has this pandemic earned our special global attention? There are three things at work.

- The pandemic is happening at a time of significant cultural, political, social and ecological upheaval

in the world; thus, the pandemic is being conflated with other pressing crises.
- The pandemic has emerged in a time of unprecedented social media saturation. Every event that occurs in the world today is being effectively viraled in sequence. Thus, the events seem more important than if they occurred just five to 10 years ago.
- The pandemic fills a gap of fear being generated by the transition period we are in. In times of the unknown, humans tend to lean into fear as a default emotion.

Please note we are not sharing this analysis to take away the critical nature of the pandemic, but to make sure we understand it is a creature of this time.

With this context in mind, it should be noted that COVID-19 provides a kind of jumping-off point for the next level of transition for humanity. The new normal.

But what does that next level of humanity look like? That becomes the next big challenge for all of humanity --- making the unseen, seen. We must be able to translate what the future looks like into a solid conversation, a plan, infrastructure development, leadership and belief big enough to take on the challenge. It would be easy to say we are in unknown waters, but that would not be quite accurate. It would be more accurate to say we're at an evolutionary jump point. A blueprint for that was given to us from 2 million years ago.

Despite what so many skeptics would have you to believe, there really is hope for us, but this hope isn't a nice feeling or a Kum-by-ya moment. It is an actionable item that needs our immediate attention or the value of this new normal moment will escape us.

The COVID-19 wake up call couldn't have better timing. We're at a moment of other significant challenges that are asking us to examine the ways of the past so that we don't bring the worst of those ideas and practices into the future. Some of those practices include:

- Ignoring the urgency of climate change at the expense of life on earth.
- Food insecurity caused by wasteful management and distribution structures.
- Growing dramatic income inequality.
- Entrenched gender inequality and discrimination.
- Systemic racism that says some are more valuable than others.
- A 6th mass extinction of animal species.
- A global anxiety and depression epidemic.
- Water rights and scarcity issues.
- The constant threat of war.

These items have been around for so long we've become comfortable with them, like white noise or wall paper. But what if they're not natural occurrences but symptoms of bad approaches, bad ideology, and bad execution? As difficult as the COVID-19 pandemic is, it provides us with an opportunity to take a good hard look at how we've arranged life on earth, and perhaps, with a new set of approaches, take a new, more inspired and empowered road into the future of a new normal.

CHAPTER 2

Socioeconomic spillovers shaping the new normal.

Spillover is a socioeconomic event that occurs in a specific context, but whose extensive influence, overflows and affects another context. For example, odors from a nearby plant are negative spillover effects upon its neighbors; the beauty of a homeowner's flower garden is a positive spillover effect upon home values in the neighborhood.

There will be socioeconomic spillovers from the old paradigm into the new normal.

This chapter will show that if we think these spillovers through, we'll see that spillovers are indispensable assets and could form a unique winning formula, if utilized appropriately.

HOW ARE WE EXPECTED TO DEAL WITH SPILLOVERS IN THE NEW NORMAL?

The spillovers are just as important as technological development, new infrastructure and innovative governance models. We'll cover those things, but want to concentrate our focus on how your personal life will be affected. We'll look at how our current spillovers directly influence our quality of life decisions we'll need to make over the next few years.

Once the new normal spillovers are identified, successful transformations plans need to be linked to the resources, but using an abundant thinking mindset. Government, national leaders and activists can use this information to set national plans and priorities.

In order to create 'live new normal models' we need mainly three prerequisites, as shown in Figure (1). We need first new normal spillovers transformative change plans that are clear and simple. Then we need to appreciate the type of new normal spillovers influence suitable for our communities. And, finally, we need to experience this new model in the global cities and communities.

Figure (1) How Spillovers would shape the new normal

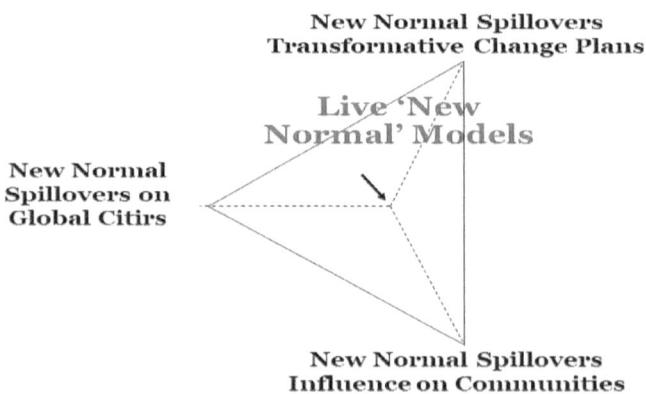

Here are major spillovers that are and will affect the new normal.

ᛥ Climate change spillover.

The single most driving factor of future life on earth right now is climate change. This man-made event of increasing world temperatures due to our civilization's CO_2 emissions is on track to change the very fundamentals of the modern world. Some of the sub-spillovers include:

Climate change refugees -- Masses of people moving from one part of the world to another to escape extreme climate conditions.

Climate change gentrification -- Where a large number of the rich and well to do move from their residents affected by climate change to other areas and, often, displacing the poor and working class who already live there.

Food security -- Extreme droughts and floods will change the regular time tables of weather we've come to expect around the world. There will be food shortages.

The climate change spillover into the new norm is that it provides us incentive to re-think how we live. Truth be told, humans rarely heed a call to action when everything is going well. It is when crisis occurs that we think more creatively, with more innovation and courage than most other times. The very fear of a drought-stricken, fire-driven and food shortage planet is leading us to ask the most important questions of our lifetime:

- ✓ Can we save life on earth?
- ✓ Are there better options on how to live?
- ✓ What tech innovations can we employ to change the course of civilization?

It can be argued that this level of self-evaluation for humanity is needed and necessary for life to thrive on earth. It provides an opportunity to grow beyond given ideas of the past. Any society or civilization that does not re-invent itself in a time of crisis is either dead or dying. This is an opportunity to come alive. The new normal incorporates this level of thinking, not has a crisis-based occasional event, but as a standard way we consider things.

, KNOWLEDGE GAP SPILLOVER.

There is a gap between new jobs being created in the world and the number of people who have developed the skills to do those new jobs. The tech knowledge and intellectual understanding required for these jobs continues to increase each year. At the same time, fewer and fewer people have accumulated those skills. This spillover not only is a big driver behind the growing income inequality gulf between the well-to-do and the working class, but it also is a factor in countries now experiencing population bust and a falling tax base. If there are fewer people in your country and fewer people in the world who can keep up with the knowledge gap of new jobs, more communities will find themselves economically "upside down" with an older population that needs more services but with a shrinking tax base.

The upside to this knowledge scarcity is that it will help the community to think about the application of education in the new normal and perhaps creative ways to empower those who have not accumulated knowledge wealth. Universal Basic Income (provide a monthly income distribution to each family, no questions asked) is an example of an opportunity to better fortify our communities.

‣ TECHNOLOGICAL SPILLOVER.

As companies struggle to fill jobs that fewer and fewer people qualify for, they will increasingly turn to the technology to fill the gap. Advanced AI is making this a logical next step. This development is a double edge sword. While AI will make our lives more efficient, it will displace more and more people *since the focus of AI is the challenge the very concept of job.* What do we do with displace worked due to technology? The other part of this spillover is that AI increases the efficiency of the system and dramatically reduces waste. Applied to the food distribution system, this waste could turn into surplus in food deserts and neighborhoods around the world.

‣ HUMAN CAPITAL AND EDUCATION SPILLOVER.

The standard government approach toward job loss due to technology and the education gap is to send displaced workers to education and retraining. The reality is that retraining has not had a particularly successful track record when it comes to integrating those retrained individuals back into the work force. However, it can introduce retrained individuals to discover parts of themselves that lay hidden for years. It will open them up to new possibilities they would not have discovered otherwise. These new possibilities can be transferred into new ways of making a living. In the new normal, the emphasis will be on finding and creating opportunities rather than seeking a job.

› THE COVID-19 PANDEMIC SPILLOVER.

While it may seem that the COVID-19 pandemic started the shift and an evaluation of the spillovers, the truth is that this moment in time is a culmination of events at least a century in the making. However, the pandemic has brought a level of attention to practices that have been unsustainable or just a result of bad judgement. Here are some of the spillovers….

A future world of periodic and intermittent quarantines and lock downs --- During the pandemic, the one thing that seemed to "flatten the curve" of infections is large scale quarantines and lock downs. The other benefits from this action? Cleaner air, reduced carbon emissions, emerging wildlife, even more family time. This kind of spillover means that intermittent quarantines and lock downs have to be considered as a way to manage future pandemics as well as provide a chance to catch our breath in a race to slow down climate change.

The full display of the wealth gap -- The wealth gap or global income inequality has been a storm brewing since the late 1970s. The COVID-19 pandemic allowed us to see more demonstrable and graphic displays of this disparity as the poor, working class and communities of color around the world were caught in the cross hairs of this global crisis. This, while the well-to-do were able to access health care above and beyond what others were able to enjoy. This spotlight on something hidden in plain sight will make the new normal a time to address it in policy, laws and social activism.

Acceptance --- While the spillover from the COVID-19 pandemic has created levels of stress and anxiety in communities around the world, it has also led to a certain degree of acceptance about what some would believe as unacceptable in the modern world. This level of acceptance spillover could work well in helping us finally deal with longstanding issues.

HOW THE SPILLOVERS WILL AFFECT YOUR COMMUNITY IN THE NEW NORMAL.

Studies today emphasize that creating community empowerment and socioeconomic impact are becoming significant predictors of resident happiness, quality of life and successful business development in a region, especially in the long term.

When we speak of communities in this book, we're speaking of relatively smaller groups of people (compared to cities) who have bonded together either geographically or through other common factors for the greater good of their members. This means communities can exist in cities, rural areas or suburban spaces.

Using spillover developments and effects as catalyzing calls to action can be quite effective in one of the true modern sources of participatory government --- communities.

Communities have three distinct advantages in taking spillover effects and turning them into opportunities:

Communities are not too big. Large organizations can be too intimidating to foster participation with all of its members.

Communities can be networked more efficiently. Again, size is a key point. It is not unreasonable to believe that things can be too big to be managed effectively. Smaller, networked groups can be more agile in a time of new challenges and opportunities. Networking allows them to access the success and wisdom of other communities.

Communities still allows us to connect. In 2013 scientist Matthew Lieberman released a study that says one of our primary drivers is to connect. Connection happens more fluidly in smaller communities as opposed to larger, more socially challenged cities. This connection makes it easier for us to find common ground and solutions when unexpected changes produce challenges.

THE IIEP, COMMUNITIES AND SPILLOVERS.

One of the main projects that the authors have been involved with is the International Inspiration Economy Project (IIEP). IIEP have been established in 2015 in Slovenia with the purpose of spreading labs and practices that would prepare the socioeconomies to the new normal, as introduced by the COVID-19 pandemic.

IMPORTANT SOCIOECONOMIC PROJECT IMPACT AND SPILLOVERS FRAMEWORK.

In order to get a clearer in-field experience of communities moving to the new normal, Dr. Buheji will share three cases in which he was directly involved.

Case study of a community transitioning to the new normal.

The case studies focus on amplifying the three types of IIEP projects that targeted to tackle issues relevant to poverty elimination, health care services improvement and women development which are highly expected during the new normal transformation stages. Each of these three cases was carried in different countries: Bosnia, Bahrain and Mauritania.

First Project- Poverty reduction through re-Inventing humanitarian organizations role in the community

Summary on the Socioeconomic Project

Many developing countries still suffer from the challenges of poverty elimination despite many government and non-government services. Bosnia and Herzegovina (B&H) are one of the countries where poverty elimination have been facing many challenges since the civil war ended in 1993. The need to reduce the effect of poverty was the focus of many humanitarian

NGOs in B&H. Therefore, a project started with an NGO called Merhamet to transform their performance goals from poverty alleviation to poverty elimination. Buheji (2019a, b, e).

To understand the problem in proper perspective, the Merhamet beneficiaries in the city of Bihac were analyzed to see whether they represent the city's poverty population. The demographics of the beneficiaries were collected against their different assets capacities and their functionality, i.e. how much these beneficiaries can they self-dependent.

A social assessment for all the cases of families getting support from Merhamet was collected, categorized and then codified. A thorough review after carrying out random sampling shown that families need to be re-assessed again according to more precise criteria. Then a table was established to help detect the priority weight matrix that would measure the special demographics of the different poverty cases. Low-income families who received two services or more were checked and socially assessed again. For example, the reasons for providing cooked food for each family were re-evaluated. Cases of the families and the individuals in need were categorized as per their age eligibility and functionality. For example, from 60 - 75 years= green, i.e. most eligible for support. While 59 - 45 years= yellow, which means have a high probability of being either turned to be out of the waiting list if fit to be trained for self-sufficiency. The rest of ages of 44 – 30 years = red, 29 years and below too, which means that individuals should not receive help (or should receive temporary assistance).

In order to make each person live with dignity and be fully independent a specific amount was considered as per the following: For a single person = US $35 and for a whole family of 4 = US $150, per week. The first step towards a practical solution was to get youth, from the families 'in need' and cases supported by Merhamet, to get involved in the management of the NGO services. Then a plan was set to building a network that ensures the interaction between those youths and the youths from the donating families.

The 'lower priority' applicants were removed from the waiting list. The observation forms were set for collecting a fresh collection of the socioeconomic status data of the families who receive more than one service (i.e. the upper threshold). Criterion such as: gender, marital status, age, ability and functionality, diseases, government support, support from other NGOs, family support, homelessness, financial situation, duration of support from the NGO, number of children/dependents, type of humanitarian services received, transport, were all measured with weightage for each family currently in the Merhamet support program. The purpose was to define which families are in red and yellow codes that need to be prepared to be out of the list as they are competent enough to be independent and create a social and economic contribution.

The "green" cases were finally identified, i.e. those of families proven to be in poverty, in order to reduce their number. The cases on the waiting list were re-examined, and a selection for more families in need as per the weight was admitted to the beneficiaries approved list. Those not in priority for exiting, i.e. those coded as yellow or red cases, were registered for rehabilitation and productive family programs.

Different university students and especially those of social studies college were deployed to re-study and frequently assess the NGO's cases every month, as part of an internship program. Plans were set to reduce the number of young people who receive meals from the NGO's service by 20% every year, as they have both the physical assets and functional capacity wealth that make them to contributors not receivers of humanitarian services. The target of Merhamet shifted gradually, over a period of six months, towards reducing the number of those on the waiting list, with higher priority given to those individuals who score less in their functionality. Since the waiting list carried lots of youth, entrepreneurial support services were enlisted as part of Merhamet new partnership strategy.

One of the main outcomes of this problem-solving lab is that Merhamet is more confident that it provides services according to real needs. Besides, Merhamet managed to strengthen its presence in the community by building new focused partnerships that helped in accomplishing more effectively focused services. Getting Merhamet beneficiaries gradually coded as (red) and (yellow), which are consistently removed from the waiting list helped to create a model for eliminating the causes of poverty. A development, management and operational teams were established to collaborate to ensure that these practices are sustained. Buheji (2019a, b, e).

Finally, a total reform in the business model of the humanitarian agency made Merhamet become a healthier organization and more profitable by starting a bakery. This mindset of starting an efficient cost centre for supplying fresh daily bread with the meals helped to cut cost by 20% since bread makes up 30% of the meal. The bakery targets now to become a profit centre, as would be the case of the new Merhamet building spaces which could be rented out for events. Merhamet strategic team initiated also trusts funds, that focus on helping the NGO to expand its role as a social transformation agency that targets to eliminate poverty in Bihac and be a model for B&H and Eastern Europe.

THE PROJECT IMPACT AND SPILLOVER.

Synthesis of this project shows there are main three impacts that build the outcome solution of the problem, as illustrated in Figure (2). The first impact is the target towards the elimination of poverty. This impact brought with it, as per the case, the first spillover that is the development of techniques suitable for the periodical assessment of poverty types and cases. Buheji and Ahmed (2019).

The second impact of the Merhamet case was the improvement of the demographics of the beneficiaries. This brought the second spillover that is the development of community engagement with the poverty elimination through partnership. Buheji (2019a, b, e).

The third impact of this case study is that it re-emphasized the role of humanitarian NGOs. This brought the third spillover that the advancement of the poor towards being more independent and with a focus on the functional beneficiaries.

Figure (2) Poverty Elimination Case Study Impacts and Spillovers

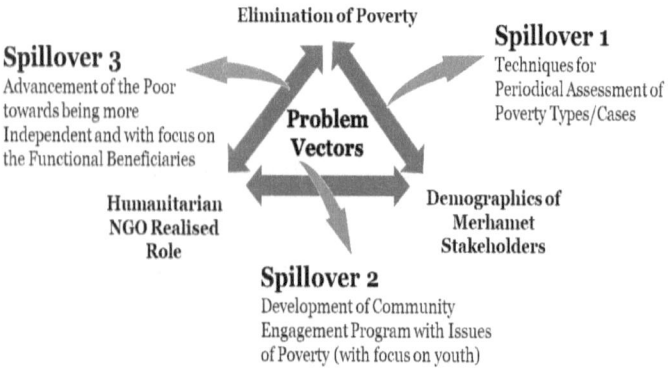

Second Project- Prioritizing Emergency Patients

Summary on the Socioeconomic Project

Many patients die every year even in developed countries due to the difficulty of being admitted as 'emergency patients' to the hospital, because of the limitations or the occupancy of the wards bed. Many hospitals do not have consistent peer reviews on the utilization and turnover of beds occupancy.

Studying the largest hospital in Bahrain, called Salmanya, revealed that the bed occupancy ratio is very high, which mean the hospital emergency bed are congested, and there is slow emergency beds turnover.

The first steps taken towards this socioeconomic issue was to study how to motivate and inspire the different medical staff involved with such a problem. A communication model was set to engage the following medical staff with the problem under study: the consultants, the residential physicians, the nurses, and the patient management services. Buheji (2019c).

The hospital, similar to all other health care organizations, found to work based on 'vertical thinking'. i.e. every physician and every department have their system for patients' discharge. The goal was to transform the hospital to work based on 'horizontal thinking'. i.e. all the departments collaborate to create a better quality of life for the emergency patients. Also, 'integrated thinking' was established in each ward, i.e. to speed up the reporting between multidisciplinary wards teams. Buheji (2019d).

The opportunities to speed up the availability of beds and to lower waiting times to receive emergency service were explored. The first opportunity was to build a model for bed turnover while increasing the level of medical and health care services provided to patients. This helped to realize the abundant time of the resident physicians, which was diverted for managing the cases to speed up their release. The methods of communication between the wards and other service departments, such as pharmacy, x-ray, labs, administration and the bed scheduling team, were improved to the benefit of assessing how emergency patients could be admitted by the freeing beds of the recovering patients.

A dashboard for monitoring of beds turnover per physician was established. The dashboard would show colour codes for a patient to be released soon as being codified as yellow, i.e. in the recovery stage. The board would have a red colour card for beds that passed the limit expected for the patient case as per the protocol of the case. A specific resident physicians' team was assigned to prepare the patient release documents on time. The same time work on developing and updating the patients' demand for beds in relevance to the type of disease protocols and the clinical pathways. This was reflected in the 'discharge planning' and home follow-ups.

Since most patients stay after 5 pm and even over weekends because the discharge plans are not ready, more focus was given towards this area. The main outcome of the problem solution is building a new culture with a new spirit that focuses on the patients' rights to receive a bed based on the urgency of the case. The solution outcome showed the role of medical staff in 'Influencing change and improving hospital conditions without the need for extra resources. The opportunities explored and utilized in the solution helped to continuously re-frame the mindset of the medical staff and reduced their resistance to change.

THE PROJECT IMPACT AND SPILLOVER

Synthesis of this project shows there are main three impacts that could build the outcome solution of the problem, as illustrated in Figure (3). The first impact is about improving the capacity for prioritizing emergency cases and the availability of beds. This would lead to a spillover relevant to the techniques and the approaches that need to found for defining the stagnant areas and activating a holistic pull system approach that shows the capacity to absorb the higher emergency patients demands in specific seasons or times. The second impact would come from the management of the patients' clinical demand while maintaining the patients' satisfaction. Here another spillover occurs in relevance to effective discharge system that ensures the patients quality of life. The third impact, in this case, is the development of the hospitality services that help to the professional management of beds with higher accuracy.

Figure (3) Emergency Patients Prioritization Project Impacts and Spillovers

Third Project- Mauritanian Women Wool Production from Rural Villages

Summary on the Socioeconomic Project

In the capital city of Nouakchott in Mauritania the women are used for manually weaving the handcrafted carpets from the camel wool. These Mauritanian women come from different village to the capital and stay for three weeks away from their village to earn a living. These women would sit before the looms and weave the rugs, in

You and the New Normal

a process that might take them as long as a year for each large carpet.

The supply of Camel Wools comes from all over Mauritania and the African Sahara Desert. The abundance of the camel wool is so much with no clear power of sales. Clearly, there is no specific style of packaging that enhance the profit margin of sales, and there are no marketing strategies. Although the manufacturing process is being completely environmentally friendly, the carpets are not marketed as an Eco-System product. All the carpets don't carry the story of the weavers, be it old or young women or those with disabilities.

The first proposed socioeconomic change was to distribute the vintage wooden loom carpet apparatus in different areas of the Mauritanian rural villages. The requirement was that there should be four women working on each apparatus. Thus, the target is to create independence opportunities for more than 200 women from different families working on fifty apparatus spread throughout the country. If the apparatus operated in two shifts, this would increase the possibilities of more production and also several people working on it, and this would reduce the production cost.

Each group of women cells were given an amount of camel wool enough to do two carpets of 3x5 meters, as a start-up loan. The factory would own the loan of the wool and also the apparatus. Each production of the carpets/rugs would be graded for quality when bought by the factory. The factory would ensure that the workers would have peer to peer development as a mobile training centre.

The marketing team would work on packaging the carpets and define European outlets that would be interested in buying this eco-friendly product. The marketing team would ensure that each carpet would have a story about: the life of the women who made the carpet, the heritage of Camel wool handcrafting in Mauritania and its differentiation, besides the guarantee from third parties.

The outcome of the project is that it enhanced the quality of Life for handcraft women and their families. More income could be generated while maintaining, working within family and village setup. The proposed outcome solution would also improve the eco-tourism in Mauritania and spread the unique brand of Mauritanian wool industry. The outcome of this problem solution is the sustainability of the uniqueness of high-quality production of hand-woven carpets industry in North-Eastern Africa.

THE PROJECT IMPACT AND SPILLOVER.

Synthesis of this project shows there are main three impacts that build the outcome solution of the problem, as illustrated in Figure (4). The first impact enhancement of the camel wool carpets from all over Mauritania. This led to the first spillover that raised the capacity of the country for using the opportunities inside the camel wool production problem for the benefit of the human condition. Buheji (2019c).

The second impact focused on maintaining an eco-friendly production while also preserving the village identity and family stability. This led to the spillover of a community-based development that exploits the opportunity of for-profit social program.

The last impact of this case is that is based on differentiating the story of each product which led to the spillover of the capacity for the differentiation of the type and price of the product through eco-friendly products that improve the profit margin of the marginalized, i.e., in this case, Mauritanian women.

Figure (4) Impacts and Spillovers for the Development of Mauritanian Women Working in Wool Carpets

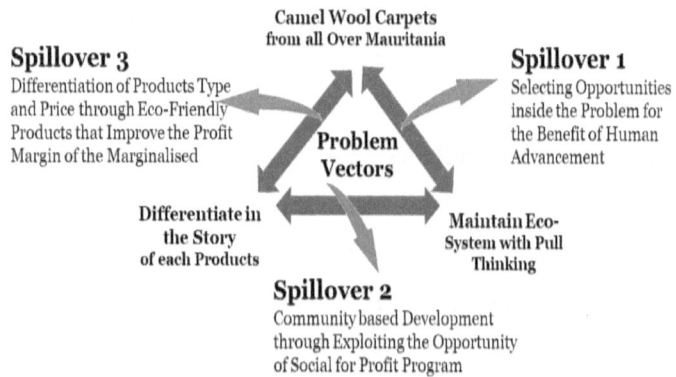

CHAPTER 3

SEVEN NEW WAYS OF THINKING IN THE NEW NORMAL.

To transform into the new normal, we need clear and applicable ideas that we can share with other people. Here are seven ideas that will make a difference in our speed of transformation if implemented in our communities, leadership models, organizations and corporations. These approaches are the bridges that would help us move from the dysfunction of the old paradigm rules to implementing the new rules of this new time.

1. WE MUST DEVELOP THE POWER OF CURIOSITY.

As we mentioned earlier in this work, we live in a vastly more connected world than any of us could have imagined just a few short years ago. This is both a blessing and a challenge. The connection idea increases our possibility of success, as the early humans did 2 million years ago. This connection idea also makes us vulnerable to the same threats around the world, as demonstrated in the COVID-19 crisis. The problem we face is that nation-states still live and legislate as if our interconnected world does not truly exist. The climate change crisis is a perfect example of what happens in a world where individual needs trump the needs of the collective. At this writing, we have yet to meet the goals set to address The Paris Climate Accord CO2 emissions rules to slow down the rate of the climate crisis. This is mostly because some nation-states refuse to participate, believing that saving the planet from a long slow death works against their individual and immediate best interest. This logic is based on the idea that somehow you can be safe in one corner of your house, even though the rest of the house is on fire. Some would say this is simply humans being humans --- self-interested, vain, and self-indulgent. We would submit that this behaviour has something to do with how our nation-states organize our leadership systems.

Many systems are designed to promote self-preservation, when they should be designed to promote the power of curiosity. Curiosity is the tool that allows us to go beyond the known and ask questions of the unknown. Having the courage to drill down on essential questions frees us from the stifling effect of what happens when we get stuck on "what's in it for me?" We ignore the power of curiosity at our own peril. The COVID-19 pandemic has exposed that blind spot in our thinking.

There are three major benefits that fostering curiosity develops in societies and leadership which allow effective transformation and transition to occur more smoothly in a challenged and changing world.

One of the most powerful benefits that curiosity brings is the ability to foster openness to new experiences and new ideas. One of the big challenges many countries faced during the early part of the COVID-19 pandemic is that there was an inability to look beyond systems they already had in place to battle something that was above and beyond what they had experienced before. Curiosity about this new challenge could have opened them up to transparency, innovation and new ways of dealing with problems. In a Forbes Magazine article in March of 2017 called "What Happens When Leaders Lack Curiosity", testing showed that the single best thing that companies can do to promote diversity and inclusion is to hire leaders with high openness scores. They will not just be more willing to understand and connect with people who are different – both demographically and psychologically – but also create more diverse teams and inclusive cultures.

On the other side of that coin, when leaders lack curiosity, they will hire in their own image, creating homogeneous teams where differences are stigmatized rather than celebrated. That homogeneous is not just regulated to their hires. It also creates status quo thinking. This can be the worst possible kind of thinking in a time of dramatic change and challenge.

Another benefit of curiosity development is that a curious mindset has a higher tolerance for the unknown. If you wanted to find out what scares most people, it is the myth of the unknown. We say myth, because most people don't understand how the unknown works. Tomorrow is the unknown, yet we live as if the day is promised. Most of our existence is navigating the unknown. The myth is that some single event in the future is unknown and everything else is assured.

The curios mind navigates this space, not with a disregard for most people's tolerance of the unknown, but with a healthy respect for the unknown as a way of life. Curiosity somehow allows people who possess it to build a bridge over the waters of the unknown, understanding that there is an on-going chance they might fall in those choppy waters. They simply build the bridge anyway. Even more, the curious mind does a mix of three major streams --- data, experience and intuition. These are the same streams I and another Futurist blend to come up with a reasonable view of the road ahead. Curious minds mix these streams and have a broader capacity for answers than non-curious minds.

The curious mind also is more likely to accept their mistakes and missteps in their decision making. The non-curious have an urge for closure on their mistakes so that no one knows that they were responsible. This kind of closure is a form of self-esteem self-defence. Again, these are the kind of actions we saw entire countries take at the beginning of the COVID-19 pandemic before having to reconcile their self-defence with the obvious growing cases of the virus in their states.

Another indicator of a curious mind (but certainly not restricted to it) is that curious minds often have a broader category of friends than the non-curious mind. As mentioned earlier, about high openness scores, curious minds extend their beliefs beyond business and professional worlds and into their personal worlds. They have managed to create friendships across racial, cultural, gender and geographic lines because their curious mind won't let them get boxed in. Their friends, as well as their mental tool kit, are broader and bigger.

This sounds great for *identifying* a community and a curious leadership team, but how, if possible, can you develop curiosity?

There are seven things you can do with your team or community to bring out this highly desirable talent and trait.

Get your people out of their comfort zone. We often talk too much to the same kind of people and get the same kind of answers. Get them out. The challenges of this transition period will be beyond what traditional thinking can produce.

Create an environment of asking questions. We're not used to that. We think questioning means our organization is weak. On the contrary. Creating an environment that fosters questioning is a sign of how strong the organization is.

Listen without judgement. We have been in environments where others have judged those who asked questions because the questions didn't line up with their preconceived notion of how the question was supposed to be asked, who is supposed to ask the questions and what the questioner is supposed to look like. Often this is cultural bias disguised in a dismissive attitude toward someone who didn't grow up in your same culture or community. Train yourself to listen without judgement, even if you disagree with the premise of the question. You could be missing something incredibly important.

We don't want to paint a picture of the curious mind being above reproach. On the contrary, they are subject to the same mistakes of false assumptions, cultural bias and misinformation as the rest of us. They simply don't let those handicaps stand in their way to make courageous decisions that are bolder than the non-curious mind. The challenges of the world ahead include technological evolution, geopolitical realignment, climate change crisis, moving from a petroleum-based present to a renewable energy future and socioeconomic transition that will be unlike anything we've seen before. Curious minds are the ones strong enough to navigate this level of change.

2. WE MUST CHAMPION VISIONARY THINKING.

Vision, the ability to imagine and see possibilities beyond physical evidence, is usually left to politicians, leaders of faith and CEOs as a tool of the ambitious and well connected. The truth is, all human beings have been born with it as part of the human package. To what degree we develop it usually determines the quality of our lives. Even more, our degree of developing it will determine our readiness in managing the new challenges of the future. Here are three things you must do in developing your visionary leadership style...

Use your vision to develop trust with your colleagues. Visionary leadership does not simply mean that you can see the immediate future. It means you have the courage to speak about what can or cannot happen in the next few years and how that will affect you and your team. Think about it from this angle: Let us say you were in a conversation with your fiancé' and asked you're spouse-to-be about their future plans, what do they see for your family and how do they see themselves in the world. If they answered "I have no idea. I have not given it thought", you can be sure that you will develop a high degree of anxiety around trust in the relationship. You are marrying someone who does not seem to have a clue around things that will directly affect your quality of life. The same exists for organizations and communities. If you seek to navigate the new world emerging, you must provide an honest assessment about the risks ahead, but match it with the opportunities that also exist at this time.

Overly optimistic projections are wrong. Visionary thinking is NOT about painting a rose-coloured view of the way forward. Unfortunately, those seeking to organize their communities and organizations for the changes to come to think this is how visionary leadership is done. Its a way of placating your team into not understanding the tremendous and real risks of the journey. This is an easy way to create a lack of trust in anything you say. When I had my advertising agency years ago, I would tell my employees...there are real and sure challenges ahead, but based on the information we have today and the power of your talents, we will do well in the next year." This means I have given them my honest assessment of the risks and the opportunities this time of change presents. It also means things could change because of the dynamic nature of the time.

<u>Consult your brain trust.</u> Visionary thinking is not developed in a vacuum. It is the process of gathering data so that you can come to a solid, clear and qualified guess at the road ahead. This means consulting your friendly neighbourhood Futurist (ah-um), checking in with colleagues around the world and using your best intuition to make an assessment.

<u>Use vision as a powerful unifier.</u> We would submit that when an organization, country, city or community is in discord, the lack of a common vision is usually at the centre of the problem. Vision doesn't mean everyone agrees on an item. It means that the vision or a view of the road ahead is so big, inclusive, affirmative and compelling, people are willing to put aside their differences to make that one thing happen, first and foremost. If people are not embracing your vision, it may not be big enough.

<u>Vision makes you put in work.</u> Most people may not look at vision as a process, but it is. It requires you to flesh out a passing thought or idea. It demands that you think it through so that it's more than a wish or a dream, but an action plan. Vision requires you to make, at least, a mental commitment to something that used to be fringe or a blip on your radar. Developing the vision means you are now committed to seeing something through.

Vision is not a silver bullet to solving our problems, but it is a powerful tool, especially in a world during the transition. Anyone talking about the status quo in a time of change should raise a red flag in your mind. A time of transition is asking us to expand our capacity of what is possible. This is where true leadership starts to emerge... through vision.

3. WE MUST TEACH AND LEARN HOW TO BE ADAPTIVE.

For the record, scientist Charles Darwin *never* said this: *"It is not the strongest of the species that survives, nor the most intelligent that survives. It is the one that is most adaptable to change."*

However, Darwin's colleagues and disciples forwarded this quote as they read over his notes and looked at his evidence. The species that were able to live on when an environment rapidly changing were those who developed adaptable approaches and skill-sets. The biggest and most cutthroat species were eventually phased out. That's why we don't have dinosaurs running up and down our streets today.

As we face new challenges to our time on the earth in the form of the climate crisis and perhaps other unknown pandemics, there are campaigns being run to forward status quo thinking while Darwin, African philosopher Credo Mutwa and journalist Naomi Kline keep pointing to in their work --- adaptation is the road to future success.

Technically speaking, adaptive learning is the process of custom learning experiences that address the unique needs of an individual through just-in-time feedback, pathways, and resources (rather than providing a one-size-fits-all learning experience). For the record, we are not dealing with the technical aspect of adaptive learning, but rather, I'm speaking of learning how to be adaptive. Learning how to be adaptive requires whole life learning in real-time to manage massive change. This is the evolutionary bridge between two epochs. This is moving from one-way of how life operates to another.

Adaptation means rethinking the very processes that got us to this point of crisis in the first place and replacing them with ones that are more in sync with the current time. It means considering new, innovative and creative approaches to how we are to live going forward. It means embracing just the opposite of the ways that we've employed over the past few centuries. It means a complete re-evaluation of what works, what doesn't, and what can we do now with the tools at hand.

Both climate change and the COVID-19 epidemic gave us clues as to how we are to deal with the coming challenges. However, many countries and individuals defaulted into old behaviour (lack of transparency, deception, denial and blame). From an evolutionary perspective, this presents a challenge into evolutionary theory: if this is a time of transition for society and the world, why did these leaders exhibit *new* behaviours that would have made the challenges more manageable? I would argue two points:

- The allure and benefits of power make it difficult for those in control to open the door for new ways forward.
- Evolutionary leadership and adaptive behaviour people will emerge when old-world behaviour starts to fail and cannot adapt to the demands of a new world.

You and the New Normal

To that end, here are behaviors of leaders, organizers and visionaries that are emerging now so that they may serve better over the next few years:

Adaptive leaders embrace opportunistic thinking. They literally see possibilities when others see the danger.

Adaptive people are resourceful. They have accumulated resources along the way and are now using their cache in their leadership role.

Adaptive people are looking down the road at what could happen. These people have a knack for looking ahead. And if they cannot see down the road, they call people who can.

Adaptive people see systems when other people see challenges. Systems are the infrastructures that keep a world operating in a certain way. Adaptive people understand that infrastructures, like everything else, need to be evolving, changed or reformed to be successful in an evolving, changed or reformed environment.

Adaptive people operate from a set of values. One of the great challenges of the old world is that a sense of values --- guiding moral principles -- are absent. People were and are making decisions based on situational ethics --- that is, decisions that were based on ideas that met the personal needs of the decision-makers at that time. This conflicted with the image that a shared set of values existed between leaders and communities. Adaptive people have a tendency to restore the faith of fair play and trust as currency, recognizing the new world will rely on them.

Adaptive people keep their minds open when others are close down. In times of crisis, status quo thinkers tend to play "small ball" at a time that asks them to go big. Why? Small ball, or reducing variables and risk, *seems* to make things more manageable in a time of stress. Adaptive people see just the opposite. They believe keeping their minds open for things they did not see before is the best way to manage the tumultuous time of change that we are in. New challenges will need new ideas as opposed to small, often used ideas. Adaptive people understand this.

Learning how to be adaptive in a time of transition is not a cliff note from a motivational speech. It is literally part of our adaptive heritage as humans. Adaptive learners are often called "early adopters" because of their ability to navigate a changing environment first. As Darwin so accurately suggested, this is not about the strongest, fastest or smartest, but about those who are able to adapt to a new world.

4. FEMININE PRINCIPLE LEADERSHIP MUST BECOME FIRST NATURE TO US.

Feminine Principle Leadership is an all-encompassing idea that leadership is best when empathetic, considerate, lateral, based on shared responsibility and shared benefits. While it is being called Feminine Principle Leadership, the foundation is based on the ideas of the Southern African philosophical construct called "Ubuntu" (I exist because you exist and vice versa). The reason this leadership model is being called Feminine Principle Leadership is because it reflects the feminine tendencies in human interaction, particularly in the ways that mothers engage children and other family members. The most important questions of this leadership model are "Is everyone okay?" "Is everyone being fed?" "How can I help?" "Share your things." "Look out for one another." To be clear, feminine principle leadership is not the exclusive domain of women. Men share in it too.

How can the application of Feminine Principle Leadership better manage the coming challenges of unknown crises? How does this management model compare to contemporary best practices?

The real strength of Feminine Principle Leadership is the capacity it can provide to management. During the early part of the COVID-19 pandemic, the government of Denmark told private companies struggling with drastic measures to curb the spread of the COVID-19 virus, that they would cover 75% of employee salaries if they promised not to cut staff. This bold move did the following:

- Relatively speaking, it stabilized the business community as it navigates the unknowns of a pandemic.
- It provides clarity for workers and helps reduce anxiety.
- It provides a definitive plan while the government works on other parts of the challenge.

This reflection of Feminine Principle Leadership asks the basic question "is everyone alright?" This is not a warm and fuzzy platitude, but a concrete form of response based on a philosophical construct. The definitive nature of this response puts the government leadership in a place to make clear decisions based on caring for the greater good of the body they govern as well as provide confidence and goodwill among the people it serves. It should also be noted that Denmark has one of the highest tax rates in the world, so it is reasonable to expect a robust government

You and the New Normal

response to serving its people in a crisis. But the essence of governments is to provide support and protection for their citizens. If it is not doing this basic function, *what is it doing?*

Would the challenges of history be different if Feminine Principle Leadership were applied to past leadership crisis moments? Let us pose these questions in "What if" scenarios.

- In the March 2015 edition of Scientific American, a report was released that indicated Exxon knew, in 1981, about climate change and the challenges it would pose to the world in a few short years. They chose, instead to embark on a mission of disinformation to protect their profits potentials. In theory, if Feminine Principle Leadership was the leadership model of Exxon at that time, it is possible a different set of questions would have been asked, perhaps leading to a different 45-year decision. Questions would have revolved around how do we make sure everyone is okay, or how do we look out for one another under these circumstances or even how we can help create a new set of options from this?
- Would the French Revolution have happened at all if Marie Antoinette and her husband operated

from Feminine Principle Leadership and asked questions like "Is everyone being fed?" or "share your things" or "are the children alright?"

- Would the Herero and Nama massacre of 1904 - 1908 have happened if the German colonizers had asked not how to take land from the indigenous populations of then Southwest Africa, now Namibia but instead asked if they could create ways to share the land, their wealth and friendship. It is worth asking if a different leadership model would have saved the lives of nearly 100,000 people from being gassed to death in the first case of genocide and ethnic cleansing in the 20th century.

Feminine Principle Leadership is the direct opposite of the command and control leadership model that emphasizes control in the hands of the few. In some ways, the challenges of the day are a referendum on the command and control leadership model. Is it sustainable? Can it be transformed to become something better? Does it have a flexible capacity under duress? Is the bureaucracy created under it nimble enough to manage an immediate crisis? What has been the command and control track record? Should we engage in another model that is more in keeping with the challenges we now face? The challenge of change and transition means that models that have little room for flexibility and efficiency will become more and more problematic.

5. WE MUST LEARN HOW TO ANTICIPATE FUTURE "BLACK SWANS".

When the COVID-19 virus swept into the world, many societies were overwhelmed simply because they had not anticipated something as comprehensive as this pandemic. But if one were to read enough of the material surrounding pandemics, one could have easily predicted the crisis and, in some cases, even avoided the full brute force of it.

Black swan events are unexpected, or events not predicted that happen out of the blue and go on to shake the very foundation of an organization or society. While many organizations have emergency preparedness plans on the books, black swan events have a clever way of catching us off guard. Even more, by their very nature, they are things that we cannot prepare for. In other words, the plans on the books are only guidelines and suggestions based on the *last* black swan event as a model.

While there is a narration making the rounds that the COVID-19 pandemic was completely unexpected, the evidence says otherwise. Apparently, it had been anticipated by experts for some time. Further evidence shows that it was not only predicted but anticipated by governments around the world, especially those who had the unfortunate experience of dealing with the SARS pandemic as well as the H1N1 pandemic. With this in mind, how could so many in the world be so dramatically unprepared for something like this? The unpreparedness was complicated by an initial lack of transparency in reporting and trying to utilize a 20th-century skill set in a world that demands 21st-century approaches.

In order to navigate the world of the immediate future, we must now develop teams whose entire job is to anticipate the future and help mitigate "black swans". Allowing this level of thinking to be part of your process does not take away the risk, but helps it to become more manageable. No, you do not need a team of thousands or a giant budget to manage this level of risk. In ancient kingdoms, a chief executive needed a "seer" or someone whose top job was to look ahead and anticipate things to come.

6. WE MUST DISMANTLE THE "RICH PERSON" BIAS.

Are rich people smarter than the rest of us? The short answer to this enduring question is no. Rich people aren't inherently smarter than the rest of us, but they have more proximity to money. That is, they have more access to the tools that create material wealth in modern society. Some would argue that being in proximity to money is the thing that makes rich people smarter than us. Let's look at that....

Structural racism and class discrimination can support incompetent rich people while suppressing really smart people who do not come from a tradition of great wealth. That is not a rule, but the opportunity exists. Systems that support material wealth don't necessarily coincide with systems that support smart decision-makers. However, rich people, on average, are luckier than many other people, and there's math to prove it. Even more, that luck is enhanced if you're born in the right country that puts a high premium on material gain, with the right infrastructure and with the right social mobility tools. This throws a wrench into the idea often championed by motivational speakers that everyone has the same chance at being rich...it's just your thoughts that set you apart.

Thoughts increase the odds, but random chance makes all the difference.

Something called "Survivor Bias" can fool us into believing that if we simply follow a model set by a particular rich person, we too will be rich --- something often pitched by the latest best-selling book author or motivational speaker. Telling you to be like Bill Gates is compelling reading, but it is misleading at best. Gates happened into a certain and unique set of circumstances.

There is no support for the idea that rich people are inherently smarter, collectively, than everyone else. But there is data that shows our deference to the well-to-do increases the odds of being exposed to bad leadership. In a time of change, this bias can make a good situation bad and a bad situation worse.

7. WE MUST ADDRESS INCOME INEQUALITY AS PART OF CHANGE AND TRANSITIONAL MANAGEMENT.

Here are three reasons why change and transitional management cannot or should not be separated from the problem of income inequality.

One- The COVID-19 pandemic has exposed how vulnerable working-class families and the poor are, due to years of a widening global income inequality gulf. At this writing, it appears that those who have the least resources are the ones most negatively affected by the current pandemic. Higher death rates, more people out of work and no resources to care for loved ones are only a few of the challenges that plague working-class families when the gulf between the rich and the poor continues to widen.

Two-Income inequality has already made it difficult for the working poor to manage the day to day living, even before the mass change occurs. According to a 2018 report from the United States called "The Fourth National Climate Assessment," income inequality keeps vital resources away from communities that need them most in dealing with day to day challenges. Working poor communities are in no position to deal with the jolt of massive, systemic change unless we re-prioritize spending and access to resources.

Three- Change and transformation management for all people in a society is an issue of morality. The old adage is that you can judge the quality of a country on how well it treats their most vulnerable citizens. Massive change and transformation will put another layer of management challenge onto leaders of organizations. Their ability to properly allocate resources to meet all of their people says volumes about the values of the leadership.

Change and transformation management requires a set of skills unique for this time. Another set of skills and management tools are necessary to meet the needs of a growing underclass at the same time of managing change.

8. WE MUST MAKE RACIAL AND GENDER EQUITY OUR MISSION CRITICAL PRIORITY.

For too long we've allowed a global system to maintain racial and gender inequality established eons ago in the old paradigm. We claim that we are not participating in those wrongs, yet we regularly embrace the benefits we're provided from that same infrastructure on automatic pilot. The indigenous wisdom, ideas from a new perspective, collective support and bold leadership were all lost in the prior world due to systemic racism and bias. The new normal recognizes that the best use of all resources is to bring everyone in. Leaving some out due to personal preference is a vanity of the disturbed that will collapse any chance of hope in the new normal. Dragging the old world violence into a nascent and fragile world is the perfect recipe for disaster. Platitudes and fund raising dinners haven't worked. Establishing a new foundation will.

Many of these ideas have been shared before. However, the new normal is a rare window opening to a new possibility for our communities. This is the time to move those ideas forward.

CHAPTER 4

HOW THE NEW NORMAL WILL DEEPEN OUR RELATIONSHIPS.

Fear is tough.

It is both a powerful motivator and a liar. It can significantly distort reality when dealing with other people -- both yours and theirs. A significant amount of our energy in relationships is based around how we manage our personal fears, the fears of the other party, our collective fears and the garden variety fear that comes up in the course of life. However, the key to making relationships more deep and profound in the new normal is our ability to be vulnerable. Let's start with that agreement as we walk further into deepening relationships in the new normal.

Being quarantined with someone over extended periods of time can lead to one of two things…a breakup of the relationship or a deepening of it. At this writing, the data from the COVID-19 pandemic and the self-isolation requests from governments all over the world are pointing to the latter, at least with engaged couples. In a report released in the April 2020 edition of Business Wire magazine, it appears that engaged couples are as eager as ever to prioritize their love as they navigate having their world turned upside down through lock downs and quarantine. Although most to-be-weds feel anxious (71%), stressed (62%) and overwhelmed (50%), more than 6 in 10 couples say sheltering restrictions have strengthened their relationships. Most commonly, the positive impact on engaged couples' relationships has resulted from finding new ways to spend time together (64%), reminding them what they love about each other (64%) and discussing challenging topics (54%), from finances and unemployment to preparedness for illness and death. Their relationship health priorities include managing finances together (49%), navigating disagreements in a healthy way (31%), and focusing on sexual intimacy (34%), though nearly a third of engaged couples are having more sex currently than before the COVID-19 pandemic started in the US. All of this suggests healthier marriages in the years to come. On the other side of this hopeful direction is also the rise of domestic abuse around the world during the COVID-19 crisis. Limited movements imposed everywhere have forced couples to spend more time with each other.... this includes the dysfunctional or violent relationships as well. The New York Times published an article in April of 2020 that said there has been an 18% increase in domestic violence in countries like Spain. China and Italy aren't far behind.

Even with the ugly and brutal history of violence against women arriving as a secondary infection from the pandemic, the new normal has opened the door to the opportunity for deeper more profound relationships. The engine behind this opportunity is, as I mentioned earlier, managing fears and shared experiences. Sharing experiences has a distinct way of creating bonding between people that ordinary fast paced life does not. Sharing the experience of the COVID-19 pandemic has given people a reason to have conversations with total strangers as well as long-lost relatives. There are interesting developments happening around the world that show spending more time together in human contact either in person or virtually, is teaching us the value of human relations above all else. We are not the only ones noticing this....

Many are using personal networks, social media and MeetUp groups to sustain a sense of connectedness during these isolating times. Meeting up together online for drinks, chats, quizzes has helped to bridge the social gap and forge new connections.

The Open University, March 30, 2020

The forced isolation has served as a reminder of how much our loved ones mean to us. And with this new found importance of connection, families are coming together in ways like never before.
Thrive Global, April 18th, 2020

More people are now turning to virtual counseling as a means of managing the pressure, depression and anxiety that can come along with the COVID-19 pandemic and quarantine. "We're dealing with the mental health challenges that are out there and seeing the pain and disruption to life," Dr. Amy Cirbus Clinical Director of Talkspace said. "But, the heartening part is that people are actually reaching out."
The Rebound, May 12, 2020

To be clear, the new normal introduced by COVID-19 will tear many relationships apart due to the stress of financial challenges and personal 'ghosts' we've avoided because we've been busy earning a living. Both of these developments are not automatic death sentences. They does provide an opportunity to drill down on what's next, our support systems, our life plans, gathering our resources, developing ideas we've put off and partnering with others for surviving and thriving in a new normal.

Like the weather, the pandemic provides a common subject matter of conversation, a common personal concern level and an existential experience that suggests we're all in this together. The rise of video communication software such as Zoom, Slack and Go-To-Meeting means there are new avenues to connect with others, especially as the pandemic crisis lingers on. This level of bonding suggests that the grounds for deeper and more profound interactions in a world that's been forced to huddle together can lead to deeper more meaningful bonds, thus deeper relationships. There are five ways we can take advantage of this global experience to make it a part of our way going forward:

1. Share the latest of how you are navigating through a time of rapid, unprecedented change. Then let your communication partner share their experience.
2. Create a plan and a sense of adventure for the next few years as change, challenge and opportunity run hand in hand during this period of time.

You and the New Normal

3. It's okay to be not okay in a time of transition. Create space for your partner to share their feelings, anxieties and fears as well as their dreams and hopes.
4. Connect with organizations that are talking about a way forward right now. Chances are, they can provide an affirmative atmosphere for you to explore the new world unfolding. Chances are also high you'll find fellow travelers.
5. Use rewards and positive motivation. Give your partner something if they are willing to be vulnerable and share their 'ghosts' during quarantine. During change, the carrot approach is always more effective. The stick has a way of making a more fearful and unknown atmosphere worse.

These are very forward approaches that are at your disposal. Remember that the new normal will produce a level of fear in many people not seen before. Giving yourself and other people in your life room to talk it out will not only ease those fears, but create a level of bonding needed at this time.

CHAPTER 5

THE FUTURE OF WORK IN THE NEW NORMAL.

Over the years, we've heard economists, soothsayers and observers (myself included) speak of the coming loss of jobs due to the advancement of artificial intelligence and how robots would be the ones doing the theft. That's not quite accurate. The robots are *already* taking our jobs.

In fact, they've been taking jobs all over the world since 1980 when global companies sought new ways of maximizing profits and shareholder value by reducing their biggest expense….employees. In the United States, the favorite complaint around job loss over the past 40 years has been that corporations are outsourcing to cheap labor in Mexico, India and China. Yes, that's been part of the mix, but the vast majority of jobs in the US have been lost to robotics and their efficiency. It's just more politically expedient to point to "cultural others" as the cause.

The vast majority of lost US jobs; 88 per cent were taken by robots and other home grown factors that reduce factories' need for human labor.
Ball State University's Centre for Business and Economic Research 2016

To put a fine point on things, it's not that the robots have come to take your job. *They really are here to challenge the very concept of job.* More and more companies are understanding that you simply don't need the same amount of labor to make the same amount of money. The expansion of advanced machines is a testament to efficiency. This, of course, cheapens jobs, which allows companies to keep wages fairly stagnant as they have for the past 40 years.

The COVID-19 pandemic has put the process of mechanizing industrial and mid-management jobs on steroids. A lot of the jobs that were lost during the crisis around the world simply aren't coming back. It's important to put the job loss dilemma in context. There is a belief that people around the world have always worked jobs. The assumption is that, since the days of the cave man, people have gotten up at the crack of dawn, punched a clock and worked a 9-5. Some of this false reasoning has actually come from people who've watched the 1960s animated cartoon series "The Flintstones". The idea of

Fred Flintstone, a cave man living hundreds of thousands of years ago, getting up in the morning, going through rush hour traffic to get to Mr. Slate's quarry seemed like a reasonable facsimile of the truth for untold numbers. I hate to burst the bubble for some of you, but The Flintstones is a cartoon series, not a documentary. While we've always had work, the concept of job is only a little over a hundred years old (since the industrial revolution). So the idea of job is a fairly new one in the history of humanity. This chapter on wide spread and well paying jobs for the masses is very much a story of the 20th century.

WHAT WILL THE NEW NORMAL BRING?

The new normal brings an opportunity for us to think through the concept of job, human value and the contribution of humans to community. But first, we need to understand more about the new world we've just entered.

There is a population bust happening in the world that's gaining speed: Most, if not all Western countries, as well as Russia, Japan, South Korea, and other parts of Europe and Asia are no longer having enough children to replace the number of people dying. This is called a population bust. On the surface, this would seem like a good thing to those folks who believed our world was on the brink of disaster due to overpopulation. But, that's not the case, fewer and fewer people being born means fewer and fewer people able to work jobs, earn a living, buy a home and provides a tax base for the community.

The world population is rapidly getting older: The world is aging rapidly. There are fewer and fewer young workers in the world to support a society, particularly in the West. This group of the elderly also need more care and services than other demographics while this shortage of care workers will only get worse due to the COVID-19 pandemic.

Immigration reform around the world is in desperate need of an overhaul to meet the needs of a transforming world. Nativist protests in countries around the world to keep out "the others" is like throwing gasoline on a raging fire of an older population with fewer young people and a diminishing tax base. Communities around the world need these younger workers, but are dead set against having them. There is a deep need for immigration reform that matches the time we live in. Climate refugees due to extreme weather will further complicate these global challenges.

An aging population that is outside of its prime working years. More jobs are going away due to robotics and AI. Fewer people are being born means fewer people of working age. Immigration reform is desperately needed so that working age people can work in places that need them most. And let's not forget, we're at the beginning of a climate change crisis.

ONE ANSWER TO CONSIDER: THE REGENERATIVE ECONOMY

The new normal has provided the very foundation to re-imagine how economies work. If the past economies have generated income inequality, systemic racism, social disruption, natural world destruction, exploitation and left the world on the brink of climate catastrophe, perhaps another kind of economy that does just the opposite is exactly what's needed. I would propose The Regenerative Economy. This construct would gear all social and economic activities of a community into restoring that which was lost, while providing people with meaningful jobs of dignity. It is designed to galvanize our resources into regenerative efforts to mitigate the effects of climate change, create employment for those willing and able to work, create a reliable tax base and put teeth behind efforts to end systemic racism and gender bias. We're talking about a community that can be consciously transformed to work for the greater good through free enterprise, economic strength, a values-based belief system while meeting the current needs of people today.

This is not another take on sustainability, that is, keeping the current socioeconomic direction but finding ways of making it sustainable. This is organizing the purpose of an entire *community* to do good and do well. Here are some of the jobs that can and are being created through a Regenerative Economy effort.

Transforming empty lots: This may be particular to the US, but there are other similar programs all around the world. The Environmental Protection Agency in the US defines a brownfield as a property, once used for commercial or industrial purposes, to be used and redeveloped in a sustainable way. There are over 450,000 such properties around the country, in every community. What would happen if we created jobs and enterprise around community gardens for food security, create a renewables energy field, create an electric automobile charging station, or an enterprise development center for communities of color. These are not new ideas; we're just organizing them around an entire community.

Construction work to rehabilitate buildings: There is enough work for years in rehabilitating buildings and homes to become more energy efficient and act as a grid for distributive energy.

Paid activism: One of the great challenges communities face is that not enough of their people understand that we live in a new normal and that we can revisit ideas that seemed extreme, weird or exceptional. Paying activists to help people understand the level of change that's happening can make all the difference in developing a progressive tax base to fund this effort. The activists can use their skills in social media or through in-person events.

Disaster preparedness training: As we've discussed previously in this book, there are other challenges waiting for us in the new normal. Our ability to manage these disasters will determine how we travel through this time. This is a job that doesn't require massive tech skills, thus, they would be more accessible to older or under skillled people who find themselves out of a job because of the COVID-19 pandemic.

Techs to modernize our electric grid: The updating of our national and global power grids to reflect a new world needs to happen immediately. Technicians who've worked in the field have an opportunity to expand their capacity while making room for more technicians to join them. It' a big job that takes a lot of time, but will put people to work.

Creating recreational space and green space management: An aspect of the new normal is that it offers us an opportunity to recapture things that were lost….one of those things lost being our connection to the land. By putting money behind the effort to create more recreational space, we create opportunities for people to find space during their day for mindfulness activities. It is one thing to talk about mindfulness, but it is another thing to create spaces that support the activities. Again, this isn't a high tech job, so it leaves out fewer and fewer potential workers.

Developing regenerative entrepreneurs among disenfranchised communities: Just as we work developed spaces for entrepreneurs of many different products, we can create a workspace where communities of color and the disenfranchised can have a chance to champion products, services and tools that can re-imagine how the world works.

Regenerative agriculture consultants: Indigenous land care takers who've learned ancient techniques from their ancestors. Agriculturalists who understand data application in modern farming. Back yard gardeners who picked up a few skills along the way. Employing them as teams to help develop food security around the world can change the game.

Tree planting: The data still shows planting trees are a formidable beach head to slow down the ravages of climate change. No skill level needed for this opportunity.

Climate change weatherization: The ability of homes to withstand the extreme weather conditions brought on by climate change will be crucial in saving lives, especially as our populations continue to get older. This is a prime area for job development.

Re-evaluating the job terrain in the new normal means that we have to understand all of the factors that are reshaping the concept of job in our communities. We must also consider other economic structures that can produce jobs and work on behalf of our future at the same time. This is not an impossible task. It merely takes visionary leadership and political will.

CHAPTER 6

Re-evaluating belief in the new normal.

Belief is basically defined as trust or faith in someone or something. Belief is the driver that determines everything we say or do. When we talk about belief in the context of the new normal, we're not necessarily talking about religion, but rather the shared set of ideas that a community with diverse religions embraces in order to operate. This shared set of beliefs or values can work with different faith traditions and can work across cultural platforms, across geographic locations and across ideas. An example of a share belief would be a phrase found in almost every faith tradition "do unto others as you would have them do unto you." While "do unto others" is a well-known and shared idea, it is not the lens in which the old paradigm has operated. A more accurate philosophical belief that defined how the pre-covid world worked is: Some get. Most don't.

It may not be a formal belief, but the defector results are proof positive this is an underlying driver. From this belief, our policies, laws, infrastructure, and resources are allocated. Our belief is the lens that guides our decision making. The new normal provides an opportunity for us to re-examine that core belief and an opportunity to implement a new kind of belief that would be more in keeping with a re-imagined time of transformation. This philosophical construct of some get, most don't has a negative embedded in its premise. It is assuming a binary idea that simply isn't big enough for the moment. It's not that resources aren't finite and there aren't limits on how things are managed. But the decisions that we make from this single belief has created our worst challenges:

- Climate crisis is the result of a belief that some can use the resources of the planet at the expense of many, even as that use threatens all life on earth.
- Income inequality is the result of concentrating wealth in the hands of a few while taking away resources from the many.
- Systemic racism is the result of the belief that some more deserve that others and we can implement practices that reflect that myth.
- Food insecurity assumes that even though our current world distribution system wastes almost

half of all food produced, we don't have enough to rid the world of hungers.
- Political corruption makes peace with the idea that malfeasance, grift and theft are justified if it results in personal gain.
- Poverty is the direct manifestation of the idea that there is not enough to go around even though 10% of the population has more wealth and resources than the remaining 90%.

Most organizations that seek to make a better world have concentrated their energies on the symptoms or results of the belief system as opposed to dealing with the core issue from which all of our actions flow --- our belief or our values. Everything starts with belief.

Ubuntu is big enough for this moment.

The new normal provides an opportunity to explore and implement the characteristics of a new shared belief: We are all connected.

While this is a version of the "do unto others as you would have them do unto you" concept, a new version of this idea is necessary. The African concept of Ubuntu (we are all connected) provides the agility communities need in order to be more flexible in an environment of constant and ongoing change and challenges. Some of the benefits of an Ubuntu belief system could include the following:

- If we believe all people and all things are connected, we'll be able to creatively re-design infrastructure that sees the environment as an extension of ourselves. Policy that reflects this belief would mean new priorities around climate crisis and perhaps a more aggressive approach to climate change mitigation and adaptation.
- If we believe all people and all things are connected, more robust economic systems will be developed that will account for prevailing poverty. The focus will be on making sure our systems executes shared abundance which would reduce poverty, if not eliminate it altogether.
- Systemic racism would be seen as a direct conflict to our core belief and help us to develop policy that would address this challenge as a priority. Everything from police shootings, the prison industrial complex, global exploitation and human trafficking would get the resources needed to end their practices simply because our money follows our policies.

- Political corruption would be greatly reduced because our belief would be that theft, malfeasance and deception are morally repugnant and worthy of the highest application of the law. Once our laws reflect our new beliefs, laws will have the teeth to reinforce what we collectively known.
- Restorative justice (the idea of creating criminal punishment that helps restore the community from the wrong doing rather than exact revenge) would be seen as a way forward into a new criminal justice system.
- Food waste would be outlawed and force us to create new systems that reflect this.

Changing our core belief doesn't mean we'll experience instant success into a new normal. It only means re-evaluating a belief to see if it is collectively efficient and effective in our communities is just smart. Fortunately, the new normal has provided us with a window to do due diligence on something we may not have questioned before. If that core belief is working against our quality of life and threatening our very existence, doing a re-examination is simply prudent and wise.

CHAPTER 7

WE HAVE SOME DECISIONS TO MAKE ABOUT COMMON PRACTICES.

Here are some of the new rules that will help us move away from the systems and infrastructure that seem unable to work effectively in a new world that demands more of them.

BREAK THE BELIEF THAT SHAREHOLDER VALUE IS EVERYTHING.

Moving global corporations away from their singular fixation on maximizing shareholder value and getting them to focus on the quality of life in communities will be a step in the right direction in salvaging life on earth. When shareholder value is your obsession, it means anyone and anything can be exploited by the company to produce profit --- even at the expense of our survival as a species. In the United States, the Business Roundtable, a non-profit association based in Washington, D.C. whose members are chief executive officers of major companies, pronounced in 2019 that they were moving away from the shareholder value is everything model and moving to a more holistic approach to corporation/consumer relationship. This is a nice gesture, but if the belief of shareholders and the corporation is profit by any means necessary, this will only be a public relations effort designed to misdirect. This will require communities to hold corporations to another level of responsibility outside of PR, "giving back to the community" efforts and sponsoring another 5K run. Get them on record and leverage the power of "social media shaming" as a way to shine light on the subject. Remember, this is leveraging all tools available in a moment in time when everything is on the table.

Create progressive tax codes.

I'll use another example from the United States. From 1945 to 1975, the US saw a level of growth unparalleled in modern history. During the presidential administrations of Franklin Roosevelt, Truman, Eisenhower, Kennedy, Johnson, Nixon, Ford, and Carter, the top-tax-bracket rate was at least 70 percent, and for long periods was much more. (John Kennedy's tax-cut plan of the early 1960s took the top rate from 90 per cent down to 70 per cent.) This money went directly to empowering communities and taking care of the common good. The Nordic States of Denmark, Norway, The Netherlands and Finland have championed a higher tax rate and regularly rate in the top 10 as the best quality of life countries in the world. Of course, the application can vary around the world, but making taxes more progressive to take care of the greater good is a powerful first step.

LET'S RE-EVALUATE THE GOSPEL OF GROWTH.

Change and transformation management means everything is put on the table for a good look over. We now need to have a grown-up conversation about unlimited growth. Is it good? Is it bad? In a finite world, does it even make sense? Are there other ways of doing growth during change and transition? Director of Strategy for a global collective called The Rules, Martin Kirk asked two basic questions in a 2019 article -- "What if growth isn't as positive as you think? If we don't quickly create a new economy that isn't based on constant expansion, we will run out of Earth?" Having the conversation isn't a sin. Not having it is.

LET'S TAKE A CLOSE LOOK AT UNIVERSAL BASIC INCOME AGAIN.

A twelve-year Universal Basic Income (UBI) provided by the GiveDirectly charity in Kenya arrived at three distinct results.

A. People tend to spend free cash on necessities.
B. Contrary to prevailing myths, people in the UBI program don't work less or waste money on vices like alcohol.
C. The community where the UBI families lived actually became richer because the families spent their money in their communities.

The idea of a Universal Basic Income to families has been around for years, but the ongoing income inequality gap has helped the idea of giving people a basic income, no questions asked, has gained the attention of communities around the world. Other studies are still in the works, but the preliminary data says people don't become lazy after receiving free money….they actually become more industrious. In this dynamic time of change, opportunities abound. True leaders of their organizations, teams and communities put all possibilities on the table.

Dr Mohamed Buheji & Futurist Chet W. Sisk

WE MUST TIE CHALLENGE AND OPPORTUNITY TOGETHER.

In case it hasn't hit home yet, we are living in very interesting times. As interesting times go, both wolves live among us....the one that will drag us to hell and the other one that will lead us to a greater world. Which wolf we feed is being decided at this time. As a trend analyst, I am observing, in real-time, what decisions we're making and how it will impact the immediate future. I advocate for the feeding of the opportunity wolf, but our view of what is an opportunity and what is challenge differ widely. It becomes crucial that we understand that our mind is very adept at playing tricks on us... sometimes making everything a point of danger. There are large swaths of citizens in societies around the world right now that are actively engaged in profound disassociation with the facts in an effort to hold onto a certain set of values. You may be one of them. All of us are vulnerable.

The key to future decision making is not disassociation or pretending that the challenges do not exist. But rather, creating room for both challenge and opportunity in thinking through the challenge, then creating an answer that accounts for both. Creating an either/or zero-sum thinking process is an old-world dichotomy that sees the

You and the New Normal

world in stark, unrealistic terms. It's not this or that. It's this and that, and that, and that.

The term crisis is saddled with negativity. However, one of the formal definitions of crisis is: "The turning point of a disease when an important change takes place, indicating either recovery or death." Two kinds of worlds are emerging at the same time. The crisis asks us to make a decision on what kind of world we are choosing to embrace. Just a reminder, trends are collective decisions we've made. This means we have an opportunity to change our minds on that decision and perhaps feed the wolf that can lead to a powerful, abundant, connected and bold future. We must choose our wolves wisely.

WE MUST STEP UP THE TEST DRIVING OF ALTERNATIVE ECONOMIES.

The new world emerging is coming at a heavy cost. The economic challenges which the world has not experienced since the Great Depression are now here. The COVID-19 pandemic has exposed old economic models are not sustainable. It has also helped us understand how vulnerable we all are to problems that may seem regional or local. This vulnerability has left the door open to leveraging the global ties, but through more robust and agile models that work better. In other words, there are other economic models that are better at handling the challenges of the future than the models we may be currently using. Here are some that we may want to consider for our communities.

The Sharing Economy: This model, based on the idea of sharing technology, spaces and opportunities, is simply suggesting that sharing of resources is much more economically efficient and effective than what we have been using for the last few centuries. It uses trust as a primary currency and has developed fail-safes for trust defaults.

The Gift Economy: This newly emerging idea is social-media-based and designed to have all production donate a portion of the gross to life-sustaining institutions and problems. Its "gifts" a portion of everything we make, produce or distribute to greater social causes, like universal health care, eliminating hunger and ending poverty.

The Collaborative Commons/Zero Marginal Cost Economy: This new economic system positions itself as the successor to Capitalism in that it takes advantage of technologies of efficiency and serves the public good much better than anything we have created before. The premise is simple: we now have the technology to make basic goods and services (food, utilities, information goods, etc.) free. This stabilizes the markets and allows us to direct new resources into innovative ideas to empower society and the planet.

The Ubuntu Economy: This idea is based on the South African philosophical concept that we are all connected to each other as humans and to the greater world of animals, the environment and all living things. This economy is not necessarily based on hard infrastructure as much as it is based on the idea that all things are connected. Once that is established, the economic direction is to now create systems that understand everything is dependent on other things. This changes separation, individualism and self-serving vehicles and discourages unilateral action. It encourages collaborative, work-together models.

The Circular Economy: This economic model challenges the very foundation of the fossil energy-economic model. In this model, the manufacturers maintain ownership of the products they produce, instead of people purchasing the product individually. The manufacturer is now responsible for the maintenance, upkeep, updating and fixing of the products. Instead of the product going to the dump after we are done using it, the product goes back to the manufacturer, and the consumer is provided with a new one. The resources the old item has are now back with the manufacturer. The consumer provides a monthly or yearly fee to the manufacturer for the service, and that fee costs less than the purchase.

These are only some of the new economic models. The bottom line is that there are tons of new economic models that we now have the opportunity to pursue in this new world emerging that are asking us to expand our possibilities.

WE MUST GET COMMUNITIES TO HOST VISION PARTIES.

The great challenge that seems to show up regularly in our current leadership and social circles is that vision is seen as a kind of exotic fruit. It is "nice-to-have" occasionally for variety and to feel good, but it's never really seen as a standard tool for everyday success. Even more, the current vision of many of our leaders is on lack, fear, and scarcity. Thus, with their resources, they create political, religious, economic and social structures that reflect their vision. This is why vision parties are critical. You must develop vision parties where you and trusted friends get together over pizza, snacks, coffee, tea and talk out loud about your biggest, most empowering dreams and visions not just for you, but for your communities. Developing a vision helps people understand what may lie beyond the immediate horizon and help communicate what those possibilities and challenges may be. Merely following the status quo may even be a sign that a society has died and run out of fresh ideas to reinvent itself. Vision creates a formal infrastructure for thinking ahead. Vision development is not an exotic fruit. In a time of change, it is the lifeline between what was and what could and will be.

We must now embrace test practices instead of best practices.

We are going to step away from the term "best practices" for now and lean into another one that may be appropriate for the times. Let us list a series of "test practices" that we should be putting into action now. Some of them are tests because they have not been implemented before. The same could be said for what we now call "best practices". Someone had to test them out at some point to find which ones worked.

Let's create a regenerative economy.

We should empower our communities to lead in supporting entrepreneurship and small business development that focuses on making a pandemic ready, climate-strong community. The focus of all future grants and monies must go into supporting businesses that have a climate-strong initiative as part of their business plan. These businesses must receive priority in funding and opportunities. This creates green jobs and businesses lost during the COVID-19 global economic standstill.

LET'S MAKE OUR COMMUNITIES A MAGNET FOR GLOBAL RESOURCE PARTNERS WHO SUPPORT CLIMATE CHANGE ACTION.

Many of our communities became vulnerable cities to the current economic downturn because of our reliance on sales taxes and revenue. The unknown nature of our fiscal capacity to weather this storm means we must immediately expand our base for investors who share our vision forward.

LET'S HAVE MORE SHUT DOWNS.

No, we're not advocating you spend more time locked into your home with nowhere to go or nothing to do. However, outside of the loss of life and jobs, we saw another effect of less human activity -- less pollution and CO_2 emissions. Now that the public understands how shut downs work, we can create more periodic and well-timed shut down periods like these to keep future cases of COVID-19 in check as well as and keep reducing CO_2 emissions low. This would give us a chance to slow down the climate crisis. The remarkable rebound the environment of the earth has made during the brief economic downturn needs to be part of our ongoing mission.

GET THE FAITH COMMUNITY TO CREATE A COMMUNITY SET OF SHARED VALUES.

Malfeasance, grift, theft, dishonesty and the lack of transparency in governments around the world has created the need for a new way people can create trust. The new normal has created a window to develop this. We are moving into an era where trust is the new currency. However, trust as currency is a fiction if there isn't a set of shared values championed by that community. Just like fiat currency (the US dollar for example) a system falls apart unless there is belief that a tool has meaning. If there is not a shared set of ideas, transactions of all kind will be difficult at best, creating room for the worst of our communities to advance their agendas. A shared belief and set of values is not one that everyone agrees to. Rather, it is one that all faiths socialize in their temples, their churches, their synagogues and other places of worship.

Let's go even further. The leaders of faith can craft a set of values, but it must be carried and shared by organizational leaders throughout the community. A simple statement of values could look something like this:

Community Declaration of Values:

- We believe in the sacred value of the planet and will transform into a climate strong community.
- We believe in the value and worth of all human beings and will transform our community into a culture forward environment.
- We believe in Feminine Principle Leadership.
- We believe in reaching across cultural, geographic and ethnic spaces to connect communities here and around the world for our mutual benefit.

This is just an example. It's not that these ideas are new, but the new normal provides a reset point to implement them *differently*. Joint messaging in sermons, bumper stickers, banners and websites go a long way to bring clarity to the new normal. Remember the basics of understanding media; repeated messages affect behavior.

You and the New Normal

WE SHOULD BRING BACK VICTORY GARDENS.

During World War II, Many British and American households engaged in creating Victory gardens as a way to produce surplus food supplies for communities and cities. This should be a community campaign that helps support the food infrastructure during the crisis and minimizes food insecurity challenges throughout many parts of the world. Even more, it provides a kind of "esprit de corps" (rallying of the troops) among the general public so that we all know we're in this forward-thinking process together. Hunger has been an issue in our world for decades but was exposed and expanded even more by COVID-19 through the massive job losses.

WE SHOULD CREATE MORE PUBLIC AND EMPLOYEE OWNERSHIP AND CO-OPS.

We should leverage existing public agencies and assets (including public transit agencies, local housing authorities, public school districts, and electric co-ops), take equity stakes in companies receiving substantial direct investment (including fossil fuel) and use those funds to empower those public agencies. Let's reduce this conversation to something even more simple. Get a group of your neighbors together and create a food/clothes/things swap. You have things others want. They have what you need. Lean into your new developed relationships and create this level of co-op…either online or in your backyard.

These approaches help create the infrastructure of a new normal. Like the Ubuntu model suggests, it is not beyond our ability to create a community that works better and more efficiently than what we've embraced in the past. The key is to get these approaches in place so that the community is robust enough to endure the challengers, black swans, missteps and another pandemic. But once they are in place, they give you a foundation on which to build on.

An infrastructure built on these ideas will help make the new normal a time for re-tooling, regeneration and re-imagination.

CHAPTER 8

WHY DO WE SEEM TO BE TERRIBLE AT CHANGE?

'Transformation' and 'management of transformation', despite being a popular subject today in all type of businesses and leadership forums, seems to only get marginal results at best. In fact, a recent survey indicated that the average success rate of organizations that do well in transformation management is only around 30%. Why are we terrible at this? It may be the most important question we ask as we confront a world in fast transition due to the COVID-19 pandemic. There are obvious psychological elements at work --- fear of the unknown as well as the comfort of the familiar. However some of our transition failings may be caught in our language. Let me share some insight.

Change vs. Transformation:

Many organizations and their leaders may not understand that change and transformation really are two different approaches, and it is very important that you understand this for the new normal. Change management centres on maintaining the structure of management and building change inside of that structure. Transformation basically means everything, including structure, must be re-evaluated to make sure they are the right tools for a new environment emerging. An example would be IBM vs. Apple. IBM valued its position in the marketplace as the business computer leader. It made changes to what it saw in the marketplace as it existed and helped the IBM culture meet those changes. Apple, on the other hand, recognized that they were transforming what the term computer meant and built a company around that transformative position. Apple was transforming the world simply by seeing itself as a transformative company. Most 'change management' consultants will help you tweak around the edges, but they are not bold or visionary enough to help you look over the edge and into the new normal.

Revolutionary vs. Evolutionary:

The term revolutionary has become quite fashionable over the past few years. It is sexy, dynamic and sounds like the things leaders say at motivational gatherings and events. Images of armed men in camouflage come to our mind when leaders use this word. The problem with the term revolutionary is not only that it may not be fitting for this moment, but that is incredibly overused. From dog food to hair spray to clotheslines, we have made the revolution a novel catchphrase we use when we enjoy our mocha latte with our friends. *Evolutionary,* however, isn't suggesting rebel action, but that we take the next and natural step in a process that is begging for us to move forward. As with our ancestors 2 million years ago, circumstances have conspired for us to take that next evolutionary leap in our potential.

Status Quo vs. New Possibilities:

The great challenge for all humanity is our ability to set aside our fear and our desire to hold onto the status quo and step into the world of new possibilities that exist with transformation. The leadership that must emerge now has to be forward-thinking and opportunistic. That leadership must be able to see through the clutter of "this is what we have always done" to "can't we do better than this?" This is significant, and the subtlety of these two approaches may be lost on the uninitiated. The sheer magnitude of climate change alone means that the management, leadership and education style of the past century may have run its course. This is simply an opportunity to lead with more effective and efficient tools, ideas and philosophies.

THE HISTORIC STEPS WE TAKE.

People make history. The actions we take and the choices at this particular moment in time will shape both the immediate world and how we will be remembered by future generations. There will be those who will seek for life to go back to the way it was, but it is clear that the ship has sailed. Even more, going back to the way the world was prior to the rise of the COVID-19 pandemic means going back to climate change crises, massive income inequality and social disruption. Being able to see this moment as an evolutionary next step in human development while leveraging the new tools of a new time is an extraordinary act, and you can do it.

CHAPTER 9

THE EMERGING ROLE OF COMMUNITY IN THE NEW NORMAL.

Over and over, studies today emphasis that creating community influence and socioeconomic impact are becoming significant predictors of the reality of individual and collective success, as well as increasing the quality of life for people.

Communities are easier to manage than large scale nation states, yet well-developed communities can make a nation state stronger in the long run. Networked strong communities as shown in Figure (5) are more agile, more nimble and more flexible in a time of constant change. Even more, it reduces the odds of a small group of people making bad decisions for the entire system, and welcomes collective decision making.

Figure (5) Visual example of the transition from a central command model to a distributed, networked model.

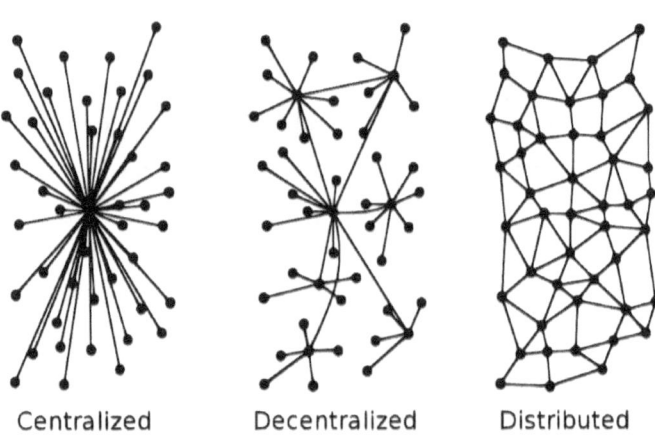

Centralized   Decentralized   Distributed

Tech types will recognize this model as the way the web has moved over the years. The centralized version of how communities and nation states operate has been shown to be at odds with a distributed web tech infrastructure on which more communities are now based on. In other words, there is a disconnect when you've implemented web-based distributed technologies to create efficiency in your community while being held hostage by an old paradigm command and control model. While a decentralized version of how a governing body can operate is better than the centralized version, it simple isn't as agile as a distributed model. Our ability to move quickly while making sure there is more participation of members to the success of the new normal.

Going back to a central point made earlier in this book about the connected nature of the new normal, communities in the distributed model will be able to have access to global resources, ideas and support while still remaining in the comfort and protection of their nation states. What we're suggesting is to create a more formal approach to this defacto development so that policy can match and support it. One of the stark and graphic demonstrations of this new emerging world was when the governor of the Maryland in the United States reached out to South Korea to obtain COVID-19 pandemic masks when those masks were not provided or available from the central/federal government in the US. This may seem like a negative development to some, but we would suggest that this is simply a scaling up of the distribution model.

Ubuntu: The philosophy of the distributed model.

There is a philosophical concept that originated out of the South Africa that matches well with the distributed community model. Ubuntu is a Nguni Bantu term meaning "humanity". It is often translated as "I am because we are", or "humanity towards others", or in Xhosa "umntu ngumntu ngabantu' but is often used in a more philosophical sense to mean "the belief in a universal bond of sharing that connects all living things."

The distributed model matches this core idea that connected networks of communities can share resources, ideas and opportunities as a successful model. This model, as shown in Figure (6), relies less on a central entity for success and more upon sharing with other communities. This is the essence of Ubuntu. Below is a model that talks about Ubuntu in even more detail as a model for the new normal. This graph goes further and shows, in detail, the other benefits a sharing and connected infrastructure can provide a community.

Figure (6) Represent the Connected Communities, referenced from Joseph Ebot Eyong (2019), published in The African Journal of Management Leadership (August).

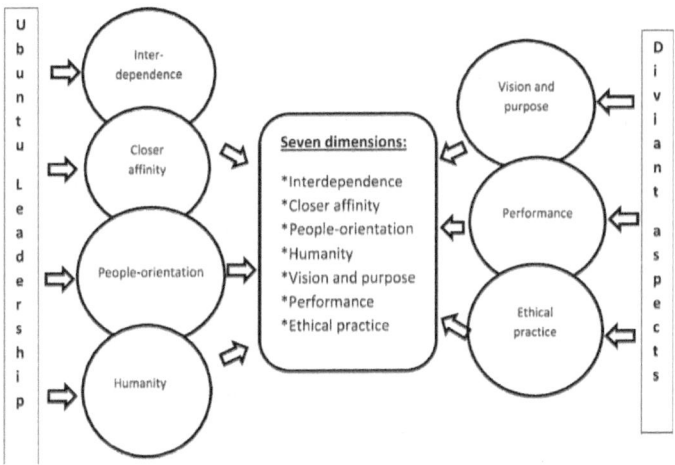

The other benefits of interdependence, closer affinity, people-orientation, an emphasis on humanity, vision and purpose as well as performance and ethical practice weave well into a new normal. Coincidentally, Ubuntu also happens to be growing in recognition and popularity around the globe.

When a community employs a socioeconomic project that embraces the distributed model and the philosophy of Ubuntu, the community has the potential to become empowered as well as empowers the network. We believe these are some of the building blocks of the new normal.

Thus, we can see a model like Figure (7) where spillovers would be considered part of the transformation journey.

Figure (7) Shows the Socio-Economic Project Impact and the Spillovers Expected

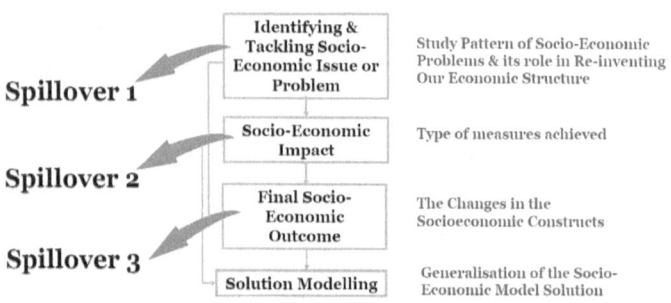

Dr. Buheji, again, shares another case study that shares his hands on experience in a distributed model of community.

CASE STUDIES WHILE TRANSFORMING TO NEW NORMAL:

The case studies focus on amplifying the three types of IIEP projects that targeted to tackle issues relevant to poverty elimination, healthcare services improvement and women development which are highly expected during the new normal transformation stages. Each of these three cases was carried in different countries: Bosnia, Bahrain and Mauritania. This give them the reliability that they can be generalised. Buheji (2018).

First Project- Poverty Elimination through Re-Inventing Humanitarian Organisations Role in the Community

1.1 Summary on the Socio-Economic Project:

Many developing countries still suffer from the challenges of poverty elimination despite many government and non-government services. Bosnia and Herzegovina (B&H) are one of the countries where poverty elimination have been facing many challenges since the civil war ended in 1993. The need to reduce the effect of poverty was the focus of many humanitarian NGOs in B&H. Therefore, a project started with an NGO called Merhamet to transform their performance goals from poverty alleviation to poverty elimination. Buheji (2019a, b, e).

To understand the problem in proper perspective, the Merhamet beneficiaries in the city of Bihac were analysed to see whether they represent the city's poverty population. The demographics of the beneficiaries were collected against their different assets capacities and their functionality, i.e. how much these beneficiaries can they self-dependent.

A social assessment for all the cases of families getting support from Merhamet was collected, categorised and then codified. A thorough review after carrying out random sampling shown that families need to be re-assessed again according to more precise criteria. Then a table was established to help detect the priority weight matrix that would measure the special demographics of the different poverty cases. Low-income families who received two services or more were checked and socially assessed again. For example, the reasons for providing cooked food for each family were re-evaluated. Cases of the families and the individuals in need were categorised as per their age eligibility and functionality. For example, from 60 - 75 years= green, i.e. most eligible for support. While 59 - 45 years= yellow, which means have a high probability of being either turned to be out of the waiting list if fit to be trained for self-sufficiency. The rest of ages of 44 – 30 years = red, 29 years and below too, which means that individuals should not receive help (or should receive temporary assistance).

In order to make each person live with dignity and be fully independent a specific amount was considered as per the following: For a single person = US $35 and for a whole family of 4 = US $150, per week. The first step towards a practical solution was to get youth, from the families 'in need' and cases supported by Merhamet, to get involved in the management of the NGO services. Then a plan was set to building a network that ensures the interaction between those youths and the youths from the donating families.

The 'lower priority' applicants were removed from the waiting list. The observation forms were set for collecting a fresh collection of the socio-economic status data of the families who receive more than one service (i.e. the upper threshold). Criterion such as: gender, marital status, age, ability and functionality, diseases, government support, support from other NGOs, family support, homelessness, financial situation, duration of support from the NGO, number of children/dependents, type of humanitarian services received, transport, were all measured with weightage for each family currently in the Merhamet support program. The purpose was to define which families are in red and yellow codes that need to be prepared to be out of the list as they are competent enough to be independent and create a social and economic contribution.

The "green" cases were finally identified, i.e. those of families proven to be in poverty, in order to reduce their number. The cases on the waiting list were re-examined, and a selection for more families in need as per the weight was admitted to the beneficiaries approved list. Those not in priority for exiting, i.e. those coded as yellow or red cases, were registered for rehabilitation and productive family programs.

Different university students and especially those of social studies college were deployed to re-study and frequently assess the NGO's cases every month, as part of an internship program. Plans were set to reduce the number of young people who receive meals from the NGO's service by 20% every year, as they have both the physical assets and functional capacity wealth that make them to contributors not receivers of humanitarian services. The target of Merhamet shifted gradually, over a period of six months, towards reducing the number of those on the waiting list, with higher priority given to those individuals who score less in their functionality. Since the waiting list carried lots of youth, entrepreneurial mentorship support services were enlisted as part of Merhamet new partnership strategy.

One of the main outcomes of this problem-solving lab is that Merhamet is more confident that it provides services according to real needs, Buheji and Ahmed (2017). Besides, Merhamet managed to strengthen its presence in the community by building new focused partnerships that helped in accomplishing more effectively focused services. Getting Merhamet beneficiaries gradually coded as (red) and (yellow), which are consistently removed from the waiting list helped to create a model for eliminating the causes of poverty. A development, management and operational teams were established to collaborate to ensure that these practices are sustained. Buheji (2019a, b, e).

Finally, a total reform in the business model of the humanitarian agency made Merhamet become a healthier organisation and more profitable by starting a bakery. This mindset of starting an efficient cost centre for supplying fresh daily bread with the meals helped to cut cost by 20% since bread makes up 30% of the meal. The bakery targets now to become a profit centre, as would be the case of the new Merhamet building spaces which could be rented out for events. Merhamet strategic team initiated also trusts funds, that focus on helping the NGO to expand its role as a social transformation agency that targets to eliminate poverty in Bihac and be a model for B&H and Eastern Europe.

1.2 THE PROJECT IMPACT AND SPILLOVER:

Synthesis of this project shows there are main three impacts that build the outcome solution of the problem, as illustrated in Figure (8). The first impact is the target towards the elimination of poverty. This impact brought with it, as per the case, the first spillover that is the development of techniques suitable for the periodical assessment of poverty types and cases. Buheji and Ahmed (2019).

The second impact of the Merhamet case was the improvement of the demographics of the beneficiaries. This brought the second spillover that is the development of community engagement with the poverty elimination through partnership. Buheji (2019a, b, e).

The third impact of this case study is that it reemphasised the role of humanitarian NGOs. This brought the third spillover that the advancement of the poor towards being more independent and with a focus on the functional beneficiaries.

Figure (8) Poverty Elimination Case Study Impacts and Spillovers Expected

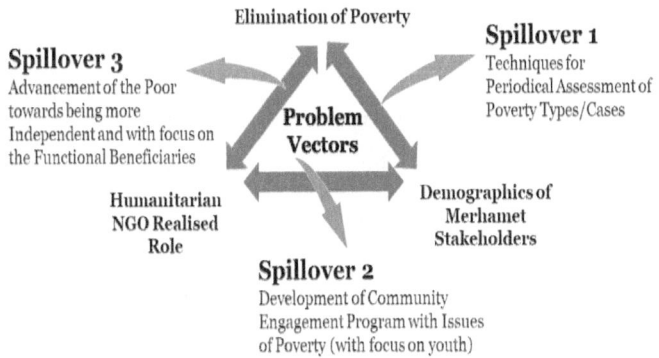

SECOND PROJECT- PRIORITISING EMERGENCY PATIENTS

2.1 SUMMARY ON THE SOCIO-ECONOMIC PROJECT

Many patients die every year even in developed countries due to the difficulty of being admitted as 'emergency patients' to the hospital, because of the limitations or the occupancy of the wards bed. Many hospitals do not have consistent peer reviews on the utilisation and turnover of beds occupancy.

Studying the largest hospital in Bahrain, called Salmanya, and revealed that the bed occupancy ratio is very high, which mean the hospital emergency bed is congested, and there is slow emergency beds turnover.

The first steps taken towards this socio-economic issue was to study how to motivate and inspire the different medical staff involved with such a problem. A communication model was set to engage the following medical staff with the problem under study: the consultants, the residential physicians, the nurses, and the patient management services. Buheji (2019c).

The hospital, similar to all other healthcare organisations, found to work based on 'vertical thinking'. i.e. every physician and every department have their system for patients' discharge. The goal was to transform the hospital to work based on 'horizontal thinking'. i.e. all the departments collaborate to create a better quality of life for the emergency patients. Also, 'integrated thinking' was established in each ward, i.e. to speed up the reporting between multidisciplinary wards teams. Buheji (2019d).

The opportunities to speed up the availability of beds and to lower waiting times to receive emergency service were explored. The first opportunity was to build a model for bed turnover while increasing the level of medical and healthcare services provided to patients. This helped to realise the abundant time of the resident physicians, which was diverted for managing the cases to speed up their release. The methods of communication between the wards and other service departments, such as pharmacy, x-ray, labs, administration and the bed scheduling team, were improved to the benefit of assessing how emergency patients could be admitted by the freeing beds of the recovering patients.

A dashboard for monitoring of beds turnover per physician was established. The dashboard would show colour codes for a patient to be released soon as being codified as yellow, i.e. in the recovery stage. The board would have a red colour card for beds that passed the limit expected for the patient case as per the protocol of the case. A specific resident physicians' team was assigned to prepare the patient release documents on time. The same time work on developing and updating the patients' demand for beds in relevance to the type of disease protocols and the clinical pathways. This was reflected in the 'discharge planning' and home follow-ups.

Since most patients stay after 5 pm and even over weekends because the discharge plans are not ready, more focus was given towards this area. The main outcome of the problem solution is building a new culture with a new spirit that focuses on the patients' rights to receive a bed based on the urgency of the case. The solution outcome showed the role of medical staff in 'Influencing change and improving hospital conditions without the need for extra resources. The opportunities explored and utilised in the solution helped to continuously reframe the mindset of the Medical Staff and reduced their resistance to change.

2.2 THE PROJECT IMPACT AND SPILLOVER:

Synthesis of this project shows there are main three impacts that could build the outcome solution of the problem, as illustrated in Figure (9). The first impact is about improving the capacity for prioritising emergency cases and the availability of beds. This would lead to a spillover relevant to the techniques and the approaches that need to found for defining the stagnant areas and activating a holistic pull system approach that shows the capacity to absorb the higher emergency patients demands in specific seasons or times. Buheji (2019c). The second impact would come from the management of the patients' clinical demand while maintaining the patients' satisfaction. Here another spillover occurs in relevance to effective discharge system that ensures the patients quality of life. The third impact, in this case, is the development of the hospitality services that help to the professional management of beds with higher accuracy. This impact would lead to a third spillover that would help to build the capacity for more reduction of morbidities and mortalities cases.

Figure (9) Emergency Patients Prioritisation Project Impacts and Spillovers Expected

Third Project- Mauritanian Women Wool Production from Rural Villages

3.1 Summary on the Socio-Economic Project

In the capital city of Nouakchott in Mauritania the women are used for manually weaving the handcrafted carpets from the camel wool. These Mauritanian women come from different village to the capital and stay for three weeks away from their village to earn a living. These women would sit before the looms and weave the rugs, in a process that might take them as long as a year for each large carpet.

The supply of Camel Wools comes from all over Mauritania and the African Sahara Desert. The abundance of the camel wool is so much with no clear power of sales. Clearly, there is no specific style of packaging that enhance the profit margin of sales, and there are no marketing strategies. Although the manufacturing process is being completely environmentally friendly, the carpets are not marketed as an Eco-System product. All the carpets don't carry the story of the weavers, be it old or young women or those with disabilities.

The first proposed socio-economic change was to distribute the vintage wooden loom carpet apparatus in different areas of the Mauritanian rural villages. The requirement was that there should be four women working on each apparatus. Thus, the target is to create independence opportunities for more than 200 women from different families working on fifty apparatus spread throughout the country. If the apparatus operated in two shifts, this would increase the possibilities of more production and also several people working on it, and this would reduce the production cost.

Each group of women cells were given an amount of camel wool enough to do two carpets of 3x5 meters, as a start-up loan. The factory would own the loan of the wool and also the apparatus. Each production of the carpets/rugs would be graded for quality when bought by the factory. The factory would ensure that the workers would have peer to peer development as a mobile training centre.

The marketing team would work on packaging the carpets and define European outlets that would be interested in buying this eco-friendly product. The marketing team would ensure that each carpet would have a story about: the life of the women who made the carpet, the heritage of Camel wool handcrafting in Mauritania and its differentiation, besides the guarantee from third parties.

The outcome of the project is that it enhanced the quality of Life for handcraft women and their families. More income could be generated while maintaining, working within family and village setup. The proposed outcome solution would also improve the eco-tourism in Mauritania and spread the unique brand of Mauritanian wool industry. The outcome of this problem solution is the sustainability of the uniqueness of high-quality production of hand-woven carpets industry in North-Eastern Africa.

3.2 The Project Impact and Spillover:

Synthesis of this project shows there are main three impacts that build the outcome solution of the problem, as illustrated in Figure (10). The first impact enhancement of the camel wool carpets from all over Mauritania. This led to the first spillover that raised the capacity of the country for using the opportunities inside the camel wool production problem for the benefit of the human condition. Buheji (2019c).

The second impact focused on maintaining an eco-friendly production while also preserving the village identity and family stability. This led to the spillover of a community-based development that exploits the opportunity of for-profit social program.

The last impact of this case is that is based on differentiating the story of each product which led to the spillover of the capacity for the differentiation of the type and price of the product through eco-friendly products that improve the profit margin of the marginalised, i.e., in this case, Mauritanian women.

Figure (10) Impacts and Spillovers for the Development of Mauritanian Women Working in Wool Carpets

CHAPTER 10

PREPARING FOR THE WAVE OF CHALLENGERS TO THE NEW NORMAL.

There will be significant challengers to the idea of a new normal. Articulating the significance of the upcoming new normal challenges starts with having a clear view of who and where we are right now. One of the popular sayings in the West regarding human behavior is that you must first acknowledge that you have a problem before you can begin the healing process. This is a truth of the new normal. It has brought us to a moment of reckoning so that we'll make better decisions for a transformed future. However, there are those who believe our current situation must be maintained because it "works for them". But that has always been the challenge of humankind over the years. The opportunity to create better ways forward exists, but there are those who simply see "new" as an impediment to their personal success. That doesn't necessarily make that person who wants to stand in the way of change a "bad" person. But you must become concerned about the person's judgement if they see the same clear, verifiable, definitive signs of distress that the rest of us see, but chooses to ignore them. If your doctor told you to eat less so that you'll lose weight, improve your heart condition and live a better quality of life, yet you refuse to eat less, the food isn't bad. Your doctor isn't bad. In fact, you are not bad. But your judgement to dismiss your doctor's advice and put yourself in peril is disturbing. The new normal is about our ability to pay attention to definitive suggestions that old ways of doing things are not just problematic but destructive to life as we know it. The new normal we find ourselves in is an opportunity to open up to new possibilities of doing things. That being said, there will be well-funded forces seeking to keep you from following your doctor's advice, and want you to make peace with your old patterns.

Potential challengers.

There are plenty of monies and powerful sources that will see the new normal time period as a way to re-establish the way things worked in the old paradigm. There are many historical examples of how well resourced sources can obscure facts...

You and the New Normal

- For years, the tobacco industry told customers that cigarettes were neither unhealthy nor addictive. The makers of Old Gold cigarettes claimed "Not a cough in a carload." And in 1994, James W. Johnston, CEO of R.J. Reynolds, told a congressional committee, "Cigarette smoking is no more 'addictive' than coffee, tea, or Twinkies." The reality, of course, is quite different. The Centres for Disease Control and Prevention estimates that five million people die every year die from cigarettes.

- In the 1930s, Joseph Stalin was determined to wipe out private farms in Ukraine and put the population into Soviet controlled communes. He instituted a policy that began starving this region. Ultimately, between two and four million people starved to death. As could be expected, the Soviet government denied any problems. What is surprising is that many western reporters repeated Moscow's interpretation of what was happening in the region.

- A report from the magazine Scientific American in October 2015 said that Exxon Knew about Climate Change almost 45 years ago. The investigation shows the oil company understood the science before it became a public issue and spent millions to promote misinformation.

The colloquial term for the method of telling you you're really not seeing what seems to be obvious is "gas lighting". It is a clear attempt to control the narrative of an important period of time. Understanding that there are vested interests in maintaining the status quo allows you to be more conscious of the influences that will seek to establish themselves.

THREE IMPORTANT THINGS TO REMEMBER:

Media saturation or can warp our sense of reality. What's important in the world today? We'll be told once we check the news. What order are they important? We'll be told that, too. What happened to the drought in southeast Asia we were told about last week? That story is out of the news cycle and suggests to us that it's been resolved or is simply not important any more. This ominimedia world may be distorting our sense of reality by regulating and spoon feeding us what is important and what's not. What's true and what's not, what a threat is, and what's not. Challengers have immense resources and have direct and ongoing influence on this media apparatus.

Repeated messages affect behavior. From advertising to political campaigns, there are those who wish to affect your behavior simply by repeating a message until it becomes part of your subconscious. Once there, it has the ability to set up as a truth, even if it is not. But if we believe it is, we tend to make decisions on that new "truth". Challengers have the resources to leverage messages to meet their particular agenda.

Follow the money. In order to find out the agenda of any organization, simply follow the money. In other words, find out where they are getting their money from and you'll know what they are seeking to do.

CHAPTER 11

A SCIENTIFIC BREAKDOWN OF HOW WE CAN MANAGE FUTURE DISEASE AND CRISES IN THE NEW NORMAL.

Many negative thoughts and panic come to the mind when we mention the epidemic or pandemic of the Coronavirus (COVID-19). It's deadly tool has rocked communities and shaken families to their core. However, as a socioeconomic and inspiring economy expert, I am trained not to forget the hidden benefits, with the momentum of the speed of the crisis, or the sudden breakout challenges. However, those that target to make an impact, or create a differentiated legacy would need to search, explore and strive to discover a method for understanding and then acting upon the obstacles that life would continuously throw at them. If one studies the life of any iconic influencer, leader, change-maker and unique entrepreneurs, we would see that they all managed to capitalize on the formula of crisis confrontation to meet their dreams. To create the desired impact under crisis, we need to keep focused and cool. Be it an individual, or an organisation, or a community, we all can thrive during the crisis, but only when they are focused and selective in our approaches.

THE IMPACT OF THE RISING FREQUENCY OF DISEASES AND CRISES.

The World Economic Forum Candeias and Morhard (2018) mentioned that even though we have passed 100 years of the deadliest epidemic in history - the 1918 Spanish Influenza outbreak, which killed around 50 million people; the frequency of such outbreaks is rapidly increasing and are expected to increase in the foreseen future.

Besides Coronavirus, many recent virus epidemic outbreaks as SARS, swine flu, MERS, Ebola, Zika and yellow fever, to name a few, have been keeping the world in the position of defence in just the last few years, Gates (2015). The World Health Organization (WHO) confirms receiving more than 5,000 early-warning disease signals per month, from across the globe in these last few years. Around 300 of these warnings are usually investigated in-depth. WEF (2018).

The number and the kind of infectious disease outbreaks have increased significantly over the past 30 years and are expected to increase more sharply in the next 30 years, The National Academies of Sciences Engineering Medicine (2016a). The economic impact of each major outbreaks as per the world bank estimated to cost more than $500 billion, or 1% of the global income. Gates (2015), Candeias and Morhard (2018).

WHAT HISTORY TELLS US ABOUT THE BENEFITS OF EPIDEMICS?

Kim and Liu (2012) mentioned how the corporate and government organizations respond to the 2009 flu pandemic opened lots of opportunities to the community. One could see this from the historical sequences over the last thousand years where most of the discoveries and inventions came in periods of crises and diseases, Caselli (2006). During such crises, societal cohesion appears, and less differentiation between race or religion occurs, Green (2011). This condition force humans to start seeing how the epidemic making them equal without gaps in income, gender, etc.

Andy Grove, the former Intel CEO, said one "Bad companies are destroyed by crisis. Good companies survive them; great companies are improved by them.", Grove (1996). Same applies to communities. Hence, we can experience and believe that epidemics brings us benefits when we choose to reject our current assumptions that cause us specific judgments. That time we would see epidemics not as obstacles but as flexible borders.

CRISIS THAT COME DUE TO EXTREME CAPITALISM.

Studies show that capitalism society would be destroyed by its own success, due to the growing hostility of the institutions of a free society and the chaos of the entrepreneurial market economy that is in the other hand is losing the sources of its intrinsic wealth. This is causing the current capitalism crisis to be more hostile every time it comes due to the extent of how the world communities frequently forget how fragile the market economy is.

The threat from capitalism practices is more realised today with the speed of the transformation of the Chinese and other similar highly populated emerging economies towards capitalism lifestyle and mindsets. The surge of travelling between countries besides the speed of life in such highly populated countries are what predicts scientists, as the Hong Kong epidemiologist Professor Gabriel Leung, have a high probability of being accurate about his estimates that 65% of the world population would be infected with the coronavirus. Caselli (2006).

Ibn-Khaldun confirms that none of the economies has continued, they all went into death. The famous cycle of Ibn-Khaldun discusses how different economies would emerge and then destroyed, although some stay for longer. (Talbi, 1981).

Capitalism requires justice to survive; this "justice" is represented today by unearned wealth. In order to stop the destruction, capitalism needs to focus on realised productivity rather than focusing on concentrating wealth with the few. Unless this move happens, we will continue to experience stories about the disasters that happen after the global risks outbreaks or disasters occurs, Gates (2015).

VISIBLE OPPORTUNITIES OF THE CORONAVIRUS BREAKOUT.

With the crisis of the coronavirus breakout, we could discover many visible opportunities. One of the first opportunities is to think again about the concepts that are dominating our life today, such as the capital economy and its resources and how fragile they are. Here we see the largest global economies such as China and Italy, collapsing in front of coronavirus. Kim and Liu (2012). Therefore, coronavirus can represent an opportunity to review how we should live in harmony in this world and keep the communication model between us without extra unnecessary boundaries. The (COVID-19) crisis is a wakeup call for those who value the quality of life and those who value financial interests regardless of the consequences to global health or human beings. These levels of values were seen clearly in some of the airlines' industry, which increased the humans' cross-infection risks, after knowing the dangers of such disease. The National Academies of Sciences Engineering Medicine (2016a).

SELF SUFFICIENT SYNERGY HUMAN COMMUNITIES.

The other visible opportunity (COVID-19) challenge brings is that it made us pay more attention to the importance of self-sufficiency and readiness, or increased preparedness. This triggers more change in the work mechanisms and to find sustainable mechanisms that help to rearrange our internal life and the way it is constructed. The synergy of humans was also clearer in this (COVID-19), which relief the tensions of many polarized areas despite the ongoing wars in many countries. Community solidarity that such deep, complex crisis would bring to the mind of the many the essence of being and living together in this world. Green (2011).

Exploiting our intrinsic power.

The other main opportunity that (COVID-19) crisis is that it would make us more aware about the 'intrinsic powers' we have within us, which are more important than other resources, including other natural resources.

More credibility and transparency in relevance to common global Issue.

One of the benefits of the shocks of this crisis is that it pushed the bar further up for the demand for better credibility and transparency in issues relevant to the common globe. The reporting and declaration of (COVID-19) disease became not only a community right, but gone further to be a human right.

Clearly, the crisis would bring in more innovative developments in relevance to medical and healthcare requirements and services. But more important the world would also more aware of the hidden enemy that we all need to fight together regardless of our differences and level of diversities.

CHINA SOLUTION – BEYOND TECHNOLOGY ADVANCEMENT.

The failure of China to eliminate the (COVID-19) crisis, despite its technological advancement, proven that we need to go back to basics when we handle complex issues. The professional handling of the crisis by China shown that technological advancement would halt or stand to be sometimes worthless, if it did not work to the benefit of humanity.

The technological advancement of China would not be any more of priority as the priority of its project one road-one built project. This is true as China discover how much its products and services need to be decentralized, if it is going to survive more coming complex crises than the (COVID-19). For example, a mega leading global Chinese company like Alibaba, can't afford anymore to be totally centralized in their products and services would be mainly dependent on operating from China. They have to diversify and distribute their network and supply chain all over the world. Coronavirus disruption has proven that we are in a phase with new entrepreneurial projects. These entrepreneurial projects focus on unlocking the capacity to reach people electronically and provide for them value-added services that reduce their fear and enhance their probability of facing hazards safely. Buheji (2019b).

Studies show that, in total, there would be more than 97 million Chinese travelling abroad by 2023, with a forecasted annual growth rate of more than 5.0% over the next ten-year, Oxford Economics (2015). Therefore, utilization of safety-driven value-added services could provide unique values to the consumers that mitigate the risks of sudden life threats.

World communication model.

Similar to SARS, (COVID-19) main challenge and still till the time of writing this chapter is about the speed of surprise that this deadly virus hit humans which required a high level of communication.

Given the increasing rate of emergence of this and similar infectious diseases and the increasing connectivity between people and the speed of the economic activity, the world communication model needs to improve. Without having a continuously evolving communication model, the underlying risks and their potential impacts would be probably increasing exponentially.

We need to understand and counter the risks across the whole spectrum of infectious diseases—from the emergence and outbreak to epidemics and, ultimately, pandemics: World Bank (2020), The National Academies of Sciences Engineering Medicine (2016a).

Although many countries managed to build good media campaigns to meet the WHO guidelines, the communication model between the critical points that could help to mitigate the risks of the coronavirus remained to be challenging.

Hung (2003) mentions about the importance of communication in the previous epidemic infectious diseases and how it is related to its level and way of spreading. Studies show that the quality of communication would define the health and hospital authorities preparedness for such disease. Thus, as Hung mentioned, any inadequate epidemiological information about the disease may hamper the speed of effective control measures application.

Studies show that both in the case of SAR and also in (COVID-19), the insufficient communication with the public led to panic. The communication between the health services within one country and different countries shown the importance of designating hospitals that would be responsible for isolation and treatments. Buheji (2016).

The learning from SARS epidemic, especially in the region of South East Asia, shows that the communication model in case of (COVID-19) can disrupt not only the health services, but can extend to the social, economic, socioeconomic, education, commerce, transportation, human rights and humanitarian services. The National Academies of Sciences Engineering Medicine (2016).

Strengthening the communication model drills could eliminate the sudden deficiencies in the coordination between the main community sectors and improve the speed of proactive and reactive reactions, or the clear command and authority at the sites which would enhance the confidence of the front-line staff and enhance their preparation.

SELF-ISOLATION FACILITIES.

The (COVID-19) epidemic shed light on basic needs for self-isolation facilities within cities, housing areas and even buildings. The isolation would prevent the cross-infection that comes from overcrowded wards and poor ventilation in many hospitals' areas. This would reduce the pressure on the medical and healthcare staff and make them more focused on the emergency cases, the intensive care facilities readiness; besides, make for them the time to work focused and to avoid unnecessary pressures. Buheji (2016).

The self-isolation approach would raise the need for health visitors, social workers, social-psychologist. Communities need to prepare a cohort of volunteers that have the capability to mitigate risks outside the hospitals' centres and even decide when cases need to be transferred to medical centres.

Maybe the self-isolation facilities would also cover facilities for the recovery phase of the patients and when they become ready to resume some of their life independence.

Hidden opportunities of coronavirus breakout.

The coronavirus crisis showed that the world is so fragile and non-resilient, but also the world is full of hidden potentials and undiscovered opportunities. Since there are new challenges that come with the complications of this deadly virus for sure, we need to look for the hidden opportunities that could boost our competitiveness in the current, new normal and the future foresighted environment. Buheji (2020a).

Hidden opportunities usually depend on the conditions of competitiveness when a crisis erupts. For example, how we could deal with the coming recession as a result of such a crisis, including managing the socioeconomic complications as a result of such a crisis.

Hidden opportunities, once discovered, can help close the 'competitiveness gap' and ensure better international and regional sustainable growth. This means the opportunities would come from the global 'diverging socioeconomic development' and the 'unbalanced growth models'.

Thus, the hidden opportunities might come from the collection of approaches that brings in chronic human issues like migration, rising non-resilience racial tensions, loss of acceptance of others, new normal and future foresight spillovers affect, fragile businesses with preventive measures, e-solutions, speed of innovation along with international communication possibilities.

CORONAVIRUS AS A MULTIDISCIPLINARY PROBLEM-SOLVING CHALLENGE.

After this disastrous incidence, we might probably cultivate a culture that lives with experimentation. No more generations that would look failure of experiments as a shame, but the real shame would be not experimenting. Hence, manage a crisis as COVID-19 pandemic we need to facilitate greater knowledge through learning by experimenting.

(COVID-19) the crisis brought the need of solving complex problems by a holistic multidisciplinary approach. The breakout needed investigation, diagnosis and treatments that used medical, technological, legal, managerial, philosophical, psychological, sociological, economical, besides historical approaches. All these multidisciplinary approaches used communication models and were applied within the organizations, the communities and the globe. Buheji (2018).

Re-thinking the business models.

The incidence of coronavirus gives us time to rethink about the business models we adopted in our life. The importance of the goodwill value in our business model should be one of the lessons learned. Government, organizations and communities could re-evaluate the ethical and the transparency issues of different industries, i.e. doing and creating businesses from things that you don't own.

It is a great opportunity for people, organizations and communities or countries to come out of the comfort zone. Here we can build new relations as we see the world after the corona challenge from a different perspective, due to the attitudes and behaviour changes that are caused by our previous assumption shock. The new changes in the ways we handle many routines in life give us opportunities to bring in new insights to life.

More attention to economics tools which lead to a better market should be mixed with better welfare, and should guarantee the equilibrium balance with profits and costs. Candeias and Morhard (2018). The sharing economy helps to emphasize how the cooperative practices as an 'entrepreneurial opportunity' and helps in better "welfare promotion".

The Corona Economy brings in Case Studies and reference for both the undergraduate and the postgraduate schools of Business, Sociology and Law Students for discussion on how Business Models are built on the basis of Cooperation and even Collaboration. Green (2011).

Realising requirements of a healthy and profitable community:

The (COVID-19) crisis in Wuhan province in China in 2020 emphasized that we need to create more resilient communities. The difference between growth and development can be seen clearly in such communities. This is exactly also what differentiates the concepts of capitalist economy and resilience economy. Buheji (2020a).

The Wuhan growth has shown that despite the accumulation of materialistic wealth, of high-rise buildings and global city services, it missed the practices of being healthy and appreciating the intrinsic power wealth. (COVID-19) the crisis showed how capitalism approaches adopted even by countries like China are causing self-destruction and created a form of individualism that uses domination techniques, like property rights, to address self-interest, with some public benefit. World Bank (2020).

The COVID-19 crisis shown the following unbalanced and unhealthy relations between the individual vs public rights, the company rights vs public rights, and the country rights vs public rights. On the other hand, China has shown very healthy professional approaches at least in dealing with the crisis as it escalates. Without persistence and perseverance, China would not have the chance to survive without huge casualties and hence the rest of the world.

(COVID-19) the crisis has shaken the greedy capital economy (i.e. doing and creating businesses from things that you don't own). Such sudden breakout tested the fragility of the attention economy we live in where the better market, better welfare could create equilibrium between profit and cost.

BUILD MORE RESILIENT COMMUNITIES.

As we are watching closely the rising impact of Coronavirus between 25th till 27th of February 2020, the news of racial and religious tensions between Hindu and Muslims in India continued to come too. The breakout of (COVID-19) crisis showed that we need to build resilient communities more than anything if we are going to survive the more sophisticated challenges.

The business continuity of both the communities and its business need to be tested again from different angles, including the interruption of the supply chain. Buheji (2020a).

Selfishness vs. Unity is a debate that is going to more debate in the post-coronavirus. I.e. Pushing coronavirus cases to outside the country or closing the borders against an alliance country are new practices that are going to affect the new normal socio-political relations too.

No country can claim or think that it can have its 'inclusive border' anymore as the world is truly becoming one village. We need to keep building resilient communities that are holistic and interdisciplinary connected. Buheji (2018).

Non-resilient communities where wealth and power are concentrated in the hands of a few and where natural resources are exploited for short-term profits would have to change the way it deals with the under-privileges otherwise it would face riots.

Being more alert for food-borne diseases.

The other livelihood side that this crisis brings is being more precise and careful about food-borne diseases. i.e. even being more particular about what to eat and how to eat any type of living creatures. The uncontrolled human appetite needs to be controlled if it creates a danger to himself and his/her fellow humans. Todd and Grieg (2015).

A study of Todd and Grieg (2015) confirms that enteric viruses are major contributors to food-borne disease, and include adenovirus, astrovirus, rotavirus, sapovirus, hepatitis A and E viruses, and norovirus. Food-borne viruses are transmitted through contaminated food, but also in combination with person-to-person contact or through environmental contamination. This is exactly what is of high probability the main cause for the outbreak of the coronavirus in the Wuhan lives central animal market in China. This place became the hub of the epicentre for this virus that is now threatening the whole world.

Todd and Grieg (2015) mentioned that such viruses coming from food could survive well in the environment, are excreted in abundance in faeces, and have a low infectious dose, all of which facilitate spread within a community. This again exactly what is now approximately confirmed about the relation of the bats faeces to this deadly disease.

Hence, there are more research opportunities in relevance to stopping the widespread of foodborne transmissions that come from cross-species infections that create more adaptation of viruses to humans. Even more precise research would be needed to see how to safely reduce the infections that would be coming from contacts between humans and animals. The National Academies of Sciences Engineering Medicine (2016b).

Appreciation for community's capacity.

When any community starts to believe that the system undervalues its capacity and know-how, they will start to be less creative. Crises as coronavirus outbreak in 2020 triggered that need for being more creative in bringing more preventions, or mitigations of risks solution, or interventions solutions to enhance the human confidence of their capacity to overcome any threats, i.e. including those of Artificial Intelligence (AI) and raise their utilisation of their profound scattered knowledge. This supports the conclusion of Yudkowsky (2008), which confirmed that AI is a threat for human beings, but humans can improve in time of crises without depending on AI.

The (COVID-19) crisis also opened a strong debate about the threat of societal population density without counting for its community capacity to deal with probability and severity formula. This crisis emphasized the importance of exploring deeper opportunities related to the sustainability of the service. Caselli (2006).

Role of the media in enhancing community capacity.

The media messages during a crisis need to be selective. Intellectual class of the society also need to be actively fit to the free-market hostility created by the Media institutions. For example, the media can take a positive role, i.e. instead of spreading the number of deaths; why not spread the number of treated cases.

Re-alignment for understanding of our livelihood.

This Corona Virus has shifted our attention and focus from being a wealth-seeking creature to more of creature that value the meaning and the value of the quality of life. While we were becoming isolated by choice and focused on the 'profit and loss' formula, we lost touch with the division of labour and its philosophy of innovation and wealth creation. This leads the curve of capitalism to sharply drops down. World Bank (2014).

The way we are dealing with a livelihood in a capital-based economy today is leading us to dump our liabilities on the youth and the new normal generation. 'Free-market economy' destroying the economic fabric of society.

As capitalism is based on the assumptions that a great a share of the world's resources could be bought, i.e. lands and natural resources could be brought regardless of who might be deprived. Finally, in a capitalist economy, we have all the opportunity to make something of ourselves.

THE CAPACITY TO LIVE LIMITLESS.

Today and due to the job market is shrinking, the bad debts of the general population are increasing. This makes the unemployed to try to make money via a mortgage. Such a mortgage crisis is only a result of the system that made it exist. The capitalism approach today to such problems is constrained by introducing bailout packages, and rising wages of workers, which amplify the problem in many developed and emerging economies with high populations like China and India. Caselli (2006).

Enhancing our curiosity economy is also another opportunity that similar incidence of the (COVID-19) crisis brings. Being curious about the sources of the problems, the probability of incidences bring is another opportunity. Buheji (2020b).

RE-ESTABLISH THE ORIGINAL CAPACITY FOR SPIRITUAL AND SOCIAL BEINGS:

We are social beings more than anything, and the incidence of the coronavirus made us appreciate more the choice of being attached to God and strengthening our spirituality. Also, the incidence awakens us of the importance of practising more social life, or of being more social-being than just humans-being. Maybe this is even more important for the youth generations who been infected with the surge of hi-tech and robotics solutions before being threatened by the coronavirus.

The crisis wakened us on appreciating the value of losing physical contact due to fear of cross-contamination. The incidence made us realise how much we love people around us and the true loyalty for those we love. This accident or incident helped to explain how we need to stick to the continual development of our social institutions so that it would impact the human action in a similar crisis in the new normal. Therefore, the (COVID-19) crisis would give rise to the importance of the social neuroscience is an interdisciplinary concept that could be devoted to understanding how our bodies would be more fit to face coming type of complex crisis through re-strengthened social ties within our surroundings.

Social neuroscience is another opportunity that we need to exploit to enhance our biological defence system and change the self-selected isolation behaviour. This would return us to be more physical social species, rather than social media and apps species. Social neuroscience would help to bring back many inter-generation dialogues and lost family values that we witness nowadays if this concept is given a chance it would help us to evolve many hormonal mechanisms that we lost in our organisms and that are essential for our survival.

Results

1 Introduction to the Research Results

The results of the synthesized literature review helped to clarify to a certain extent the socioeconomic new normal foresight after coronavirus in 2020. Therefore, the role of global cities in times of crisis, the type of consumer loyalty and reactions are visualized. The authors extract that the next 20 to 30 years would the age of the shift towards more dependence on intrinsic powers. Besides, due to the challenge of cross-infections, as per the literature reviewed, a new trend that affects all types of lifestyle industry might be triggered.

The identified research gap shows the rise of the need for epidemiology as a leading science, especially with the development of beyond (COVID-19) viruses. Then a framework that enhances our readiness for the next epidemic threat would be proposed as a result of this focused review.

2 Synthesis of Socioeconomic visible vs Hidden Opportunities of (COVID-19) Crisis

Based on the synthesis of the literature reviewed, we could list as shown in Table (2) the following visible vs hidden opportunities that coronavirus and similar crises brings to socio-economic development.

Table (2) Socio-economic Visible and Hidden Opportunities of (COVID-19) Crisis

Visible Opportunities	Hidden Opportunities
Beyond Technology Advancement	Controlling & Balancing the Growth Model
Rise of Safety-Entrepreneurship	Controlling Diverting Socio-economic Development

Understanding the Fragility of Capital Economy	Controlling Resilience Attentions
Understanding the Importance of World Harmony	Getting to Work Out of Comfort- Zone
Communication Model	Enhancing Sharing Economy
Importance of Self-Sufficiency	New Insights of livelihood and Welfare Promotion
Synergy and Community Solidarity	Eliminating the Threats of AI Domination
Intrinsic Power	Enhancing Community Capacity
Importance of Credibility and Transparency	Living with Minimalist Mindset
Value of Physical Contact	Enhancing Curiosity Economy
The Rise of Epidemiology	Re-establishing Our Spiritual & Social-Being

3 Synthesis of Socio-economic Visible vs Hidden Risks of (COVID-19) Crisis

The coronavirus and similar crises bring many risks to our socio-economic developments. The synthesis of the literature reviewed brought the following visible vs hidden risks, as shown in Table (3) that need to be either mitigated or eliminated in order for the communities to survive and safely thrive.

Table (3) Socio-economic Visible and Hidden Opportunities of (COVID-19) Crisis

Visible Risks	Hidden Risks
Rise of Cross-infections	Limitation of Human Development Capacity
Risk Inside & Outside Health Center	Being busy with AI while Pathogens are getting sophisticated
Limitation of Isolation facilities	Stagnant Business Models
Increase of Interconnectivity that Spread the Viruses	High-rise Building in Global Cities with Minimal Preparedness
Uncontrolled Human Apatite & Desire	Non-availability of Global Agreed Risk Management Framework
Free Market Hostility	Low Competence in Solving Socio-economic Complex Problems
Foodborne-Diseases	Chaos Due to Global Panic

4 Socioeconomic new normal after coronavirus 2020

Coronavirus scare lessons could manage to influence and develop our societies and changed its behaviour to think about new normal socioeconomic challenges rather than technological gadgets and designs. This means we can speed up the solutions of new normal foresight of socioeconomic challenges.

When socioeconomic patterns change, interests and power resources change along with the industrial relations and the systems that support it. The foresight that we are going to experience more instability in the new normal requires us to reassess the industrial relations and ensure that they are balanced by inclusive development that compensates for this gap.

4.1 The Role of Global Cities in Times of Crisis

Global cities are turning to be the places where the world economy is managed and controlled, and discusses the significance of economic actors and their practices in the formation of the world city network.

Global cities need to be evaluated in their resilience to similar breakouts as the (COVID-19) crisis, by understanding their level of survival, adaptation, and growth capacities, in cases of sudden or chronic stresses and/or uncounted shocks.

4.2 Consumer Loyalty and Reactions

With the spread of infectious diseases, customers loyalty started to be shaken, as the majority of those who have been influenced by the virus spread had to change their daily routines. This pandemic influenza is comparable to the disruption of climate change. World Bank (2019).

Studies port-SARS pandemic shows that most of the economic losses of such global threat do not come as a result of the disease directly, but rather due to relatively consumer reactions which starts and cascade with the failures in the economic and financial sectors, or its ability to pick-up after the crisis.

The consumers would expect that there are good healthcare measures and capacity for countermeasures in the shopping centres and consumer-related services. All the areas of the supply chain, logistics, communications, travel and mobility are expected to be ready for outbreaks.

4.3 The age of intrinsic power.

Days are proving what this author and other scientists have been calling for. What humanity needs more than ever today is not extrinsic powers, but rather intrinsic powers that come from within. The current capital economy driven mindset have limited capacity to accept this abundant choice, as it is dominated by the scarcity of need to own and control. Buheji (2019b).

If the coronavirus manages to shake up this mindset and convince humans to search about the coming solutions for human development, we would naturally see that we need to change and develop from within; if we are to develop again. This intrinsic power development might come from changing the way we deal with ourselves, or our fellow humans. This can start by eliminating threats that come from the less privileged because they cannot afford to have and eat healthy food that we eat. The other intrinsic solution might come from re-evaluating the meaning of civilisation, taking into account our materialistic lifestyle that is worthless, etc.

4.4 THE START OF NEW TRENDS.

After the settlement from the crisis, we are expected to see new trends starting in fashion, equipment, technology, services that claim or strive to be self-sustained against viruses or mitigate the impact of cross-infection. Other trends would focus on how to bring back humanity to its runway and keep it on the edge of selective advancement. However, new principles of economics would be the concern of many influencers and the motto of these leaders would be "we rise by lifting others."

4.5 THE RISE OF EPIDEMIOLOGY.

Epidemiology did not take its right as a profession that works on preventing the world and the communities from the scare of repeated killer incidents. The (COVID-19) crisis is another wakeup call after SARS and Ebola to give the epidemiologist the chance to take their role in playing as investigators for the coming crisis and be ready to react in the right time and place as one team, regardless of their nationality and positions. I think the world would be more aware of this lost opportunity this time and Epidemiology would be a job of high demand and respect, especially if they improve their case reporting with clear evidence that helps to create proper judgements, not opinions. World Bank (2014), Kucharski et. Al (2015).

Bill Gates in TED (2015) warned about the current challenge facing us today with coronavirus and urged for effective scenario planning that would improve the vaccine research besides effective preparedness training to healthcare workers. Gates (2015) reviewed the horrific global outbreak of Ebola in 2014 and mentioned how it was an early warning for what's coming.

The Epidemiologist and their co-workers might also develop new methodologies for disruptive diagnosis, not only protocol-based diagnosis, i.e. to optimise the 'Differential Diagnosis' and then reflect this disruptive approach on the treatments. The world should witness more development in evidence-based medicine as we know it today.

The epidemiologist generation needs to have a multi-disciplinary holistic mindset that studies the relation of the political, economic, social, technological, environmental and legal changes with the development of new complex viruses.

4.6 THE DEVELOPMENT OF BEYOND (COVID-19) VIRUSES.

Similar to the Spanish flu pandemic of 1918, which infected an estimated 500 million people worldwide and touched the lives of more than one-third of the world population and killed an estimated more than 33 million, we should continue to expect that virus can spread throughout the world even faster today. History Channel (2010).

It is foresighted that the next virus not only would be deadly and very difficult to diagnose in the right time, but also would be so infectious to the extent it would be carried inside our bodies while we are actively working, or travelling, or socialising without clear detectable vital signs. This means we would have a more opportunity for human-friendly disinfectants that would manage to eliminate the spread of the virus from our bodies to other friends or our communities.

4.5 FRAMEWORK THAT ENHANCES OUR READINESS FOR THE NEXT EPIDEMIC.

4.5.1 STATUS OF THE CONTEMPORARY GLOBAL RISKS FRAMEWORK.

The contemporary world needs a stronger, yet more resilient framework that addresses both global risks and explores its relevant opportunities. Currently, all the models and frameworks continue to focus on disease outbreaks that make the world resources so stretched and overwhelmed with negative energy, instead of seeing the bright side of the challenge.

The world needs to build trust in a framework that could respond with prepared capabilities that comes from the utilised opportunities and with specific resources that fit the level of risks. The contemporary poorly coordinated response in the framework is influencing the loss of lives, resources, and creating more socio-economic disruptions.

4.5.2 DEMAND FOR GLOBAL READINESS-OPPORTUNITY FRAMEWORK.

To enhance our readiness or the next human threat and also retrieve the most opportunities from (COVID-19) crisis, we need to identify the risk issues and the problems in this epidemic. This means we would need to know what to analyse, by which analytical tools and techniques followed by how to proceed.

This can be achieved by a framework that could be tested and developed in new normal and future research. The proposed framework would have phases that help to establish design thinking and see the human challenge as the coronavirus from the perspective of case design. Therefore, the framework might help us to design models that raise humanity preparedness maturity. The construct of each module would be the requirements, the specifications, the design, the implementation, the integration, and the maintenance stages.

The framework has two main factors: factor for 'risks probabilities' and factor for 'opportunities optimization', where the above modules would be continuously repeated address each one of them. The risk probabilities come from both the valid shreds of evidence (mainly primary evidence) and the conflicting data or opinions (mainly the secondary data).

You and the New Normal

The opportunities come from the problems faced in the past, or already starting to face due to this outbreak, or from the foresighted new normal. The opportunities look from the challenges of the cases reports and contradictory decisions taken. Such opportunities would help us to see the coming challenges from different perspectives and would force us to evaluate the risks with the intention of creating more proactive measures that corrective or correction measures.

The framework focuses on enhancing the powers of inference, which would help us in the new normal to resolve conflicting information about any type of human life-threating risks. The framework could trigger beneficial trends that support early diagnosis, or raise the availability of treatments at the right time. Buheji (2019c).

The most important advantage proposed by this framework is that it could trigger more curiosity currency from within and make us ask "Why?" which lead to proactive analysis and sound actions based on critical situations. Buheji (2020b).

What the New Normal Need to be Agile Again?

The world needs to be more agile and yet optimistic about the meaning and means of handling the potentially coming contagious pandemics. While this chapter has been written after less than two months since the start of the (COVID-19), where the epidemic is not yet diminished, the researchers believe that the opportunity to shore up our defences should be used. We must create a global health risk framework capable of protecting human lives and livelihoods worldwide from the threat of such infectious disease. The National Academies of Sciences Engineering Medicine (2016b).

From this chapter, we conclude that protecting the human lives and their development, as represented in their livelihoods, requires four main actions at the same time, as follows in the coming section.

Development of new normal framework

The new normal would need a framework that would manage the opportunities and optimize the constraints that the COVID-19 crisis created. The framework developed would first recognizes the scale of the risk and then defines the level of the opportunities. The relation of visible risks and the opportunities with hidden risks and opportunities, prepare us for the transformation towards the new normal. As we move to extract the hidden risks and opportunities, we should expect more complexity-based solutions that would come from a multi-disciplinary holistic mindset and approaches. This raising readiness of humanity global crisis threat and opportunities framework could have many implications on the field problem-solvers as it inter-connects risks with opportunities in three-dimensional ways.

You can see, for example, certain 'visible risks' in the new normal from the perspective that they carry 'hidden risks', besides they carry 'hidden' and 'visible' opportunity. This visualization for problem solutions can be carried for each of the four constructs representing risks and opportunities.

CHAPTER 12

RE-THINKING POVERTY IN THE NEW NORMAL.

The issue of poverty is a serious challenge to any country or community. The new normal gives us the chance to rethink our old approaches to it so that it. As mentioned in this book before everything is on the table.

Despite the expectation of great decrease of extreme poverty by 2030, as the overall global welfare level is increasing; the deep gap between the richest 10% and 90% of the rest of the world's population is expected to deepen even further due to climate crisis. As the World Bank predicts that extreme poverty will disappear by 2030, the COVID-19 pandemic is expected to wipe away any gains of the past 20 years.

Before we talk about new ways of addressing this challenge, let's take a short look at poverty over history and how it was addressed through other models before the rise of the West and economic capitalism.

Figure (11) Foresight for New Normal
Future Poverty Problems and Solutions

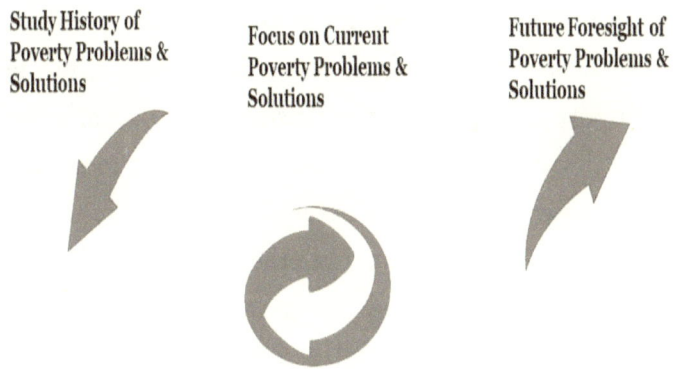

CURRENT POVERTY REDUCTION STORIES

There are poverty reduction stories that are being told around the world. Extreme poverty (living on $1.90 cents per day) overall around the world has diminished from 42% of the world's population in 1981 to 9% in 2018. That dramatic reduction is largely attributed to China which leveraged it's population to produce cheap goods through cheap labor. However, there are lots of doubts about whether the same methodologies followed for such poverty reduction would be suitable for new normal future types of poverty. For example, take the case of poverty elimination efforts done in China. Their national policies led to the annual economic growth of 7-10%. The same was replicated to a certain extent with the many south Asian countries as India and Indonesia. The massive success of the South Koreans in reducing their poverty came from their distribution system. Some countries created success stories of poverty reduction through the empowerment of small-scale farmers as India, Latin America and Brazil. The other type of well-known success story comes from programs that address the poverty causes, as the Bangladesh program for reducing women inequality with the women-focused microcredit financing without collaterals.

These success stories were suitable for the eradicating extreme poverty of the past, where poverty community were always together and could be easily identified. However, the new normal promises a future of rapid and unpredictable constant change. The automation of work environments, social distancing where more people will work online and potential regular quarantines to fight new pandemics will rock their fragile infrastructure. If this proves to be accurate, this scenario is expected to push poverty from being a complex issue to being more of a complicated problem that needs to be codified, classified and then stratified with solutions for more specific cases and based on detailed demographic data or matrix.

WHAT DO POVERTY SOLUTIONS IN THE NEW NORMAL LOOK LIKE?

Poverty in the new normal would emerge from the usual suspects -- illiteracy and negligence. Moreover, there are many reasons for poverty beyond the individual and communities such as the management of the capital by a small group of people. If both types of problems would stay in the new normal future, we should expect similar types of poverty. However, we expect the new normal future socio-economic problems would come from different practices.

If we are to think of new normal poverty solutions, we have to go into the roots of how such a problem would be occurring and think of pulling its causality in totally new ways. The concept of push and pull thinking in problem-solving is highly advised here. A push-thinking might be based on complexity theory, thus offering new tools for thinking about the new normal, while a pull-thinking might be practical efforts to solve the problems presented by changes in the condition of change, by inventing new approaches and frameworks. Any convenient socio-economic solution requires an explicit awareness of unique anticipatory systems associated with the logic of poverty.

Dr Mohamed Buheji & Futurist Chet W. Sisk

LEARNING FROM HISTORY OF HUMAN POVERTY

Poverty was attached for much of human history and throughout the entire world. The distant past had a standard of living that was very low in comparison to modern times, and this took 1000 years to change.

Global GDP before 1800 was very low. However, by early 1900, prosperity had started to increase, and by the 1950's it starts in many places around the world. For the last few decades, we have better data on global poverty (from the World Bank).

Besides the religions, the significant change in dealing with the poor in Europe started to be more precise with the Elizabethan Poor Law of 1601. The Elizabethan poor law of 1601 required each parish to select two overseers of the poor. It was the job of the overseer to set a poor tax for his or her parish based on need and collect money from landowners. The Elizabethan poor law, as codified in 1597–98, were administered through parish overseers, who provided relief for the aged, the sick, and the infant poor, as well as work for the non-disabled in workhouses.

Poverty elimination solutions in religions- Islam as an example.

Islam has a shorter-term solution for poverty as part of the welfare share system. Islam emphasises transferring a small amount of the wealth of the 'high-income group' to the 'low-income group', through the obligation of the Zakat. The Zakat target to get rid of the poor people desperation and thus to reduce income inequality.

In order to ensure equality in income distribution, Islam has set direct and indirect measures to overcome the huge gap in wealth. Ahmad (2004) mentioned how Quran praises those who can afford to give, only when they generously donate without hesitation. The promise of the reword of giving is ten times its value and flourishment of the livelihood. All Muslims who are not poor must share 2.5% of their wealth with poor people. This obligation should help to mitigate the harmful effects of income inequality. Kuzudisli (2017).

Zakat is not only an obligation from the rich to the poor, but also considered as an activity that enhances the legitimacy of wealth. Another type of obligation is the Waqef, where Muslims who can afford to grant some their amount of property or assets to the benefit of the poor or those in need for such asset. It is like sharing economy.

These deeds should not be done as if it is a mercy to the poor, but is the right of poor people on rich people.

SCALE AND IMPACT OF NEW NORMAL TYPE OF POVERTY.

Poverty is a threat; it is a source to many kinds of criminal and terrorist organizations in the new normal. In order to create balance in protecting the poor from the social and economic instability and income differences, we need to prevent the opening of greater gap of income inequality in the COVID-affected era.

In order to understand the type of poverty in the new normal besides its scale and impact, we need to appreciate the human wisdom and energy that each person carries regardless of his income, education and wellness.

Despite global progress in reducing poverty, the world is still in need to understand the types of poverty that are emerging. In the new normal the structures of poverty might come in different severe forms. I.e. to move measurement of average income and go more specific about what is happening to specific people, regions and communities.

Therefore, codification and classification of poverty in the new normal should go into the details of demographics, i.e. one could expect poverty more in the ageing group those living above 70 years or more, where their constrained choices of life would enhance their feeling of living in poverty. Also, the poverty scale in the new normal could be related to the degree of freedom, rights, intellectual property and equality.

The new normal type of poverty would be expected to come from the rapid increase of urbanisation and globalisation. For example, Non-Communicable Diseases (NCDs), like obesity, diabetes, blood pressure, cancer and cardiovascular are rapidly increasing in many developed and emerging economy countries and are bringing in new causes and types of poverty in relevant to wellness and quality of life, besides low productivity. OECD (2013). All these types of challenges are creating new risks for different types of new normal poverty that we need to mitigate.

CAN WE TRUST BIG DATA PROVIDES ABOUT POVERTY IN THE WORLD?

One of the challenges of today and the new normal is the accuracy of relative poverty data. The OECD, World Bank and IMF have now big data for more than 134 countries. This data is claimed to be collected at least once every two years, but the focus was more on absolute poverty. The accuracy of these big data depends on the parameters it is defined to measure. However, the data is not enough to create clear visibility as the frequency is becoming slower than the frequency and depth of the effective decision-making information needed.

The new normal poverty data accuracy is measured based on non-capital economy-based measurements, i.e. shifting from a measurement of depending on US $ 1.5 / a day measurement to measurement of the ethnicity, the household composition and the specific region of the country where a type of poverty accumulates. This helps us to be more ready to understand new normal poverty in different ways. This makes us more realise what might be the different new normal sources that would trap people into deprivation.

CRITICAL STRATEGIES THAT WE MUST CONSIDER.

Unless we design and implement a realistic new strategy for disaster prevention or mitigation, poverty problems might grow worse so rapidly and may reach a tipping point that would lead to an irreversible reduction in the quality of life of the majority of people in the world.

Hence, we need to anticipate and plan for the full spectrum of what type of poverty we would face in the new normal, so that we can build the capacity of the community for better monitoring and mitigation of poverty risks. Miller (2018).

The World Bank was established in 1944 with a focused goal to help countries to generate more income. Even though people living in extreme poverty reduced from 1.9 billion to 1.4 billion, this was based on the World Bank's definition, i.e. those living on less than a daily consumption of US$1.25. However, the roles of the poverty game are also changing, with more extremists, nationalists, coming up and more human-made and natural disasters are occurring with repeated trends.

Lately, the World Bank predicted that climate change would drag more than 100 million into extreme poverty by 2030. This means sudden environmental disasters, unbearable temperatures and rising water levels might drive millions of people of their homes. Hence, we might have another type of causes of poverty called 'climate refugees. Other studies confirm that poverty would increase if the IMF and World Bank continue to emphasise their measure on GDP growth, regardless of the majority 'standards of living'.

The Choudevsky work (2003) showed how the dollar metric is misleading. Displaced refugees, out of work youth, non-quality of life standards are all non-dollar metrics, yet they are highly influencing on new normal poverty. We need to visualise what structures of poverty we need to overcome in the new normal. This means we need to Imagine how poor people in the new normal, living in the compound of the less fortunate. Then we need to imagine what multi-deprivation they would need to overcome in order to meet the new life demands. This visualised new normal multi-deprivation needs to be tackled from now as it is seen and observed from those who may come out of extreme poverty but still suffer from these deprivations.

THE CORE REASON FOR POVERTY.

Poverty occurs when we fail to address the needs of: people, community and environmental conditions. These constructs can be repeated in any era, specifically with the new normal conditions of post-COVID-19. Poverty happens when we have instability in our set of habitats, livelihoods and social constructs or when experience unprepared recurring events as earthquakes or wars. Therefore, we expect new normal poverty to occur when we have an intersection between these three sources: people, community and the non-readiness of the recurring events.

Other sources of new normal poverty are linked to ignorance, apathy, disciplinary boundaries and unwise decisions. I.e. we would rarely see poverty due to chronic

You and the New Normal

hunger and primary health care needs. We would see more of human poverty where the risk of migration is very high due to conflicts or terrorism. There might be an increase in morbidity and mortality, or homelessness due to economic losses from sudden and recurring hazards.

Due to the World Bank role in addressing the current sources of poverty, elimination of extreme poverty is possible more than ever and ensuring the welfare share by 2030 is more realistic. On the contrary, extreme poverty has decreased from 1.90 billion to 720 million. Despite these pleasing developments, the number of extremely poor people in the world is still high and concentrated in specific regions such as the south of the African continent and South Asia (Cruz, Foster, Quillin, & Schellekens, 2015).

The solution suggested for the elimination of extreme poverty ensures income per capita to increase a bit more. According to the assumption, poverty will decrease, as income per capita increases, and will eventually disappear.

If rich countries started to believe that getting people out of poverty, starting with their community, is not a moral agenda only, but rather an agenda that would ensure sustainable development for the concerned parties; we won't see new poverty expanding in the new normal. Recent contemporary history tells us that countries as Germany and Turkey, accepted millions of refugees, yet their GDP has grown much more than in other previous years.

ANOTHER WAY TO THINK ABOUT POVERTY.

Before the COVID-19 pandemic, the GDP and income per capita was increasing all over the world approximately, except in the Sub-Saharan Countries. Those with high-income levels used to get the most significant share from this development. Apparently, people whose daily income is not even a dollar have to wait for the further increase in incomes of those with high income to be able to keep alive. This fact brings in new philosophy about the possibilities of new normal poverty.

Equal distribution of income between households should be the main measure for non-poverty even in the new normal. As poverty becomes widespread, this creates a negative situation for all society. The status of the poorest 10% constitutes the portion of the poverty problem that should be resolved first and most quickly.

The new normal poverty is foresighted based on people will not be independent of the society they are living in. The society needs a series of corrections in favour of the new normal poor people, at specific intervals, if the free market economy is to stay. The fact that a poor person dies out of hunger would be in the new normal the responsibility of the globe, because new normal communities would be even closer, in communication at least, to each other. Buheji (2019b)

Maybe we should change our business and entrepreneurial models.

Professor Muhammad Yunus (2018) specified that old ways of addressing inequality through charitable efforts and government programs, cannot eradicate truly the poverty problem in the new normal. Yunus saw that we could solve new normal poverty only through actions that break away from traditional capitalist mindset.

The reflection of Yunus focused on how poverty as a socio-economic problem needs in the new normal to focus on eliminating the causes of poverty that start with greater wealth concentration in the hands of the few. The blame should go to something beyond interpreting human nature from an only capitalist economy perspective. Poverty solutions of the new normal, as per Yunus, should start by changing the assumptions of having them as labour working for others to have the poor as entrepreneurs, supporting each other. This shift on the thinking framework that Yunus (2018) proposed, build new economic thinking and help in addressing the poverty of the new normal. Buheji (2019b), Buheji (2018a).

The adverse experiences that the world gone through hundreds of years should help us foresight the type of new normal poverty more clearly. New normal poverty for most would not be created by the poor, but created by the systems, and hence it would be very logical to expect it to stay but with different shape and types. Buheji (2019c).

Poverty laboratories.

The setup of the poverty labs is helping the world to know many things about poverty like what projects that need to be tackled, partnerships needed in the field before even suggesting generic policies that address the SDG1 main goal. However, all these labs do not address the new normal, it but instead focus on measuring ways to eradicate current types of poverty.

In order to build a road map for getting the new normal poor out of poverty, we need to assess their type of knowledge of what they know and what they need to know, besides their social condition. This requires action-learning research that brings in inter-disciplinary methodologies and solutions. Besides this, we need to understand how we use the new normal, in order to set up suitable new normal Poverty labs, called here for short FPL, Buheji (2019a).

FPL targets to anticipate discover and analyse the attributes of new normal poverty. FPL imagines and foresight the new normal poverty causes, practices and types and then start anticipatory activities that create basic models that make managing the new normal poverty more feasible and achievable. FPL, in short, tries to answer "what is the future, and how do we use it for the benefit of less poverty?" Miller (2018).

Narayan et al. (2000) and their team carried a type of FPL where they gathered views, experiences, and aspirations of more than 60,000 poor men and women from 60 countries. The work was undertaken for the World Development Report (2000-2001) on the theme of poverty and development. The work was titled 'Crying Out for Change' brings together the voices of over 20,000 poor from a survey conducted in 1999 in 23 countries. The other work, titled 'Can Anyone Hear Us?' also brings together with the voices of over 40,000 poor people from 50 countries from studies conducted in the 1990s. Narayan et al. (2000).

The Voices of the Poor project is different from all other large-scale poverty studies and helps to visualise what type of poverty, or poverty elimination related projects and practices we need to focus on. Using participatory and qualitative research methods, the study presents how do the poor today and in the new normal view poverty and wellbeing? Narayan et al. (2000).

Relative poverty in the new normal.

In the past, there has been much debate on absolute and overall poverty, but now the focus is shifting to 'relative poverty', as poverty is highly influenced by the place and time, we live in. Therefore, we need to foresight what is the relative poverty in new normal and in 20 years from now, so that we visualize what we need to participate in the society in which we will live in the new normal. The foresight of relative poverty in the new normal helps to set in mind the minimum standard of living to which everyone should be entitled to have to be above poverty in 30 years from now. Miller (2018).

In order to visualize new normal possible relative poverty measures, we need to appreciate the other poverty measures used today. Besides, absolute and relative poverty, there are other techniques followed by different governments or international organizations to estimate poverty. For example, the UK government measure poverty line for those below 60% of median income. Another alternative approach to defining poverty is to

You and the New Normal

look at the level of deprivation and what is the standards of living. Peter Townsend (1979), argues that deprivation should not be seen only in terms of material deprivation but also in the social exclusion from 'the ordinary patterns, customs and activities' of society. This applies of course to new normal deprivation too.

Part of relative poverty set by Townsend (1979) is the lack of resources that influence choices of lifestyle. Then we can reflect that new normal poverty threshold can be identified by the lack of necessities set by new normal standards and where the poor cannot afford it, i.e. not by choice. Therefore, one could expect that new normal poverty comes from multiple deprivations in relevance to new normal standards.

The 1983 Breadline Britain Survey and the 1999 Poverty and Social Exclusion (PSE) Survey developed the understanding of the influence of 'social exclusion on the poor'. The PSE covered the social relations, the labour market and service exclusion. The 'consensual' method, helped to develop the understanding of social exclusion and formed the basis of the current poverty and social exclusion research. The PSE approach could also be utilised for a more detailed understanding of the new normal levels of deprivation.

Dr Mohamed Buheji & Futurist Chet W. Sisk

THE MEANING OF DEPRIVATION AND POVERTY IN NEW NORMAL.

Deprivation is defined as the lack of material benefits, considered to be necessities, in a society. Hence deprivation is about living standards which are a direct measure of poverty throughout history. Measures of deprivation are not the same as measures of income; they relate to how people live. Hence, one could say that 'new normal deprivation' could be defined as the consequences of lacking new normal income and other resources, which cumulatively can be seen as living in relative poverty.

Using Townsend (1979) relative deprivation approach, one could say that new normal poverty is about lack the resources to obtain the types of a new normal diet and about the inability to participate in new normal activities. Townsend (1979) developed sixty indicators of the population's 'style of living' could also be still suitable for new normal poverty. Even though his survey was carried in the UK in 1968/69, the indicators can still be used for new normal poverty: diet, clothing, fuel and light, home amenities, housing and housing facilities, the immediate environment of the home, the general conditions and security of work, family support, recreation, education, health and social relations.

Narayan et al. (1999) seen that changing poor people's lives for the better would continue to be inherently complex, because the lack of one thing never causes poverty. Poverty be it now or in the new normal would involve many interrelated elements, and the analysis reveals that without shifts in power relations, poor people cannot access or shape the resources aimed to assist them. Any poverty elimination strategy as per Narayan work needs consider four critical elements: (1) start with the poverty realities, (2) invest in the organisational capacity of the poor, (3) change social norms, and (4) support development entrepreneurs. Buheji (2019c).

Such a study should help to explore the opportunities for enhancing the development strategies and ensure that it reaches the poor.

Eliminating poverty in the new normal starts with knowing who the poor are.

The common and false narrative around the world is that the vast majority of people in poverty are working aged men. In fact, the vast majority of the planet's poor are women and children, often ethnic minorities in their country, left to fend for themselves. Even more, poverty in developed countries is spreading silently in both rural and urban settings as women and children without means end up living with a family member or friend. They are not technically defined as poor. These codified groups need specifically targeted projects that are carried out efficiently with stratified microscale models. OECD (2013).

Thus, in order to bring those behind in the new normal we need more 'smart aid' than just 'good aid'. i.e. we need aid that helps to create stratified model solutions for the specific type of complicated new normal poverty than just an aid that targets to deal with complex poverty. This proposed framework of 'smart aid' not only it would address the chronic types of those still left behind in extreme poverty, but it would also be an excellent remedy for unpreventable wars or security crisis that occurs now and then in different places around the world. Buheji (2019b). 'Smart aid' allows transformation change in a specific sector, or area, or a group that is lagging due to new normal situation. Hence, based on the composition of the expected new normal poverty, we could design the type of intervention input. Hence, 'smart aid' would focus on 'bottom-up' empowerment approach rather than 'top-down' centralised approach. 'Smart aid' would deal with new normal poverty as an outcome, not as a status, i.e. a 'smart aid' would focus on eliminating women inequality that is leading to new normal poverty and new normal poverty-related issues as infants' mortality.

Changing the new normal poor mindset

Poverty today is addressed from a particular mindset with particular social norms. The power of social norms mindset, like the way there are still what is called the untouchables in countries as India, need to be continually challenged and eliminated to avoid such poverty sources continue in the new normal in different means.

A popular mindset about the poor is that they are not great decision makers, thus, they end up in poverty. However, researcher Linda Tirado, author of Hand to Mouth: Living in Bootstrap America found that being poor forces you to live in a permanent now. Long term planning with investments or job security or higher education is simply not available to the working poor and those in poverty. They are forced to spend all of their meager earnings as well as their waking hours on how to live day to day. Changing the mindset of what we believe about the poor in relevance to the poor today and the new normal is not simple, but it can be done.

Dr Mohamed Buheji & Futurist Chet W. Sisk

NEW NORMAL POVERTY ACTION RESEARCH.

New normal poverty action research is about exploring the progressive problem of poverty and its possible solutions. This progressive exploration can follow either participatory or practical research styles. With action research, we can improve the new normal poverty-related strategies, practices and the knowledge of the environments that lead to such poverty in the new normal. In this research environment, designers and stakeholders, along with researchers working with each other to propose a new course of action to help their community improve their work practices.

New normal poverty drove action research could follow an interactive inquiry process that balances between poverty problem-solving actions implemented in a collaborative context with data-driven collaborative analysis to understand the underlying causes of poverty and enables new normal predictions, Buheji (2018a), Reason and Bradbury (2001).

With action research balance between the researcher's agenda and the poor needs are addressed during the establishment and testing of the model solution. The research would be motivated by the goal attainment and the societal transformation that target to challenge the traditional social variables.

In new normal poverty action, research knowledge about the poor would continue to be collected through observations, as shown in Figure (12). These observations are then used to build the poverty elimination model. Then evidence on the proposed model solution would be collected and reflected on the next model. The research validity targeted would try to answer 'how to develop genuinely well-informed actions' that brings solutions to new normal poverty. We need this to continue measuring 'relative poverty', which means not necessarily measure those living in low or middle-income countries, but also those living with the same proportion in the high-income countries.

Figure (12) Poverty Elimination New Normal Action Research

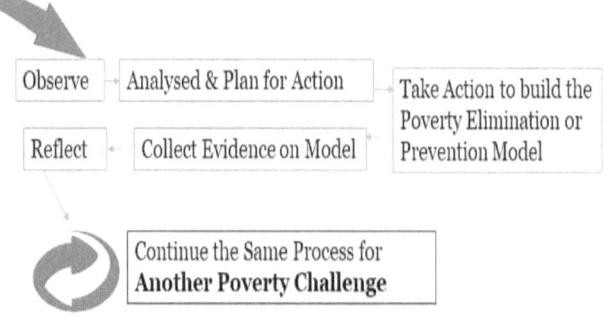

CHANGING THE POOR SOCIAL NORMS.

When we view the world from the perspectives of the poor people, five findings stand out. First, poor people definitions of poverty include economic well-being, vulnerability, powerlessness, the shame of dependency and social isolation. Degree of dependency or autonomy emerges in many countries as a classification criterion of poverty. Poor people do not talk much about income but focus on the range of assets they use in coping and in overcoming shocks.

While poverty measures to today focus primarily on the consumption and the expenditures; education and health care would become the most important dimensions of poverty. Being poor would mean that there is still a concern about insecure livelihoods and vulnerability. Narayan et al. (1999).

Most of the poor who are not involved in agriculture find their livelihoods in the informal sector. Yet most government and international attention are focused on formal employment opportunities. Poor people at the lower end of the informal sector lack any protection.

In order to come up with new normal poverty solutions, we need to understand what is the shared desirable behaviour of poor, i.e. the poor social norms. These norms can come from visualising the possible poor people interactions with landlords, traders, moneylenders and government officials. It might understand also specific norms as what women might encounter within the household.

NEW PROBLEMS. NEW SOLUTIONS.

Sustainable development goals (SDGs) managed to bring to the world high interest in statistics and projects focused on dealing with extreme poverty as a socio-economic problem. Extreme poverty was collectively tackled from the stage when it was a problem in above 50% of the world population in the early 1980s and continued till today where it reached to having on below 15% in 2018 can be categorised to be in extreme poverty. The current trend of poverty elimination efforts shows that the world would continue to build equity in all sectors in the communities. This means that more international partnerships would be done to eliminate poverty through the integration of science and public policy.

In order to understand poverty in the new normal, we need to understand what the facts of poverty are? Based on these facts, we need to study the possible positive or negative outcomes of poverty elimination and what window of opportunities need to be used. This means we need to find the right partners while gathering and analysing the facts, besides finding more selflessness driven problem solvers. Buheji (2018a).

Riel Miller (2018) seen that people use the new normal to search for better ways to achieve sustainability, inclusiveness, prosperity, well-being and peace. This is exactly why we call for the foresight of new normal poverty.

The way the new normal is understood and used is changing in almost all domains, from social science to daily life. The foresight of the type of new normal poverty can be visualised and anticipated today. The form the new normal takes in the present is anticipation.

To effectively foresight new normal poverty, we need to confront the concepts trapped in our mindset about today's type of poverty. This means we need to overturn the old frames and create new types of poverty frames. In such circumstances, we need to understand and overcome the poverty problems in the new normal.

CHAPTER 13

GLOBAL RISKS IN THE NEW NORMAL.

This chapter discusses the types of the coming and the new normal crisis and risks, with a focus on the contemporary ones. Our quality of life will be determined on by how we view and manage these risks. This more scientific look at the potential threats is part of the International Inspiration Economy Project (IIEP), which uses a methodology called Inspiration Labs (ILs) to solve current challenges in real time with "boots on the ground" work groups. In this study, we look at two times of potential challenges: prevention of crisis (called PC) and mitigation of crisis (called MC). Dr. Buheji provides a detailed look at these risks in this study.

TYPES OF SOCIOECONOMIC RISKS.

We look at a wide array of information of a community to see how systems in the community would be negatively (or perhaps positively) impacted by new crises (ie: a new pandemic, extreme weather due to climate change, etc). Communities will respond differently in the new normal. Our job is to share what that different would look like. For example, any negative impacts on human welfare such as a disruption of essential services of (food, water, energy, transport, communications, etc.) would affect the psychology of community members and would then need new tools to help in mitigation (mindfulness techniques, constant communication from community leaders, a set plan to tap global resourcest, etc.)

Through conducting risks evaluation of challenge and then comparing the results with new normal criteria, we can determine the magnitude of tolerance that would improve the social values. The way that we respond to structural shifts in global geopolitical, economic, social, technological and environmental conditions must change in order to be effective in the new world with its new tools and approaches.

The key drivers of new global risks and impacts are as likely to be unexpected events, even "black swans" that come out of nowhere. These potential scenarios must be analysed through a complex adaptive system, and that needs to be cross-boundary, and cross-disciplinary.

Since 2013 the world has been witnessing slow socioeconomic growth in almost all regions of the world. While the world economy was expected to grow by 3.9% in 2018 and 2019, the trade tensions between the US and China and the human-made disasters reduced this expectation. The migration from and to countries such as the situation in Argentina and Turkey, and the turmoil in many auto industries in India and Germany, along with tightening economic development in China, are creating an even deeper socioeconomic crisis.

The tightening of the economic situation is causing economic weakening in communities around the world, where support and supply industries are finding fewer and fewer opportunities. The great challenge is that in a new normal, old rules of how to respond to such crises, simply don't apply.

A lot of this has everything to do with the advancement of new technologies. The speed of their implementation means that infrastructure of distribution, finance, and creation have changed. Many industries are set up in a 1950s model and simply cannot change fast enough to fit into a different world. The COVID-19 pandemic opening the door for the new normal has accelerated this shift and will simply clear out those who cannot handle this level of a shift, overnight.

OVERNIGHT CHAOS.

Many world experts have been warning about the economic crisis coming in 2020, but the literature is scarce about the coming social and economic challenges in the years to come, despite the technological and general materialistic quality of life achievements. Overnight chaos, a quick and immediate jolt to the socioeconomic well being of a community, can and will happen as the new normal starts to take hold. In 2009, Greece suffered a wide-range of socioeconomic challenges that led to sharp rise in unemployment rates, precarious work regimes, rapid increases in poverty level, dramatic increases in the number of uninsured citizens, substantial income loss, widened income inequality, exacerbation of the demographic problem, disruption of social cohesion, political instability, and migration and refugee issues.

In particular, unemployment rates rocketed in Greece from 7.8% in 2008 to 24.9% in 2015 and 23.1% in December 2016 (Eurostat, 2017). The Greek poverty community rose from 28.1% in 2008 to 36% in 2014 and 35.7% in 2015. The Greek, middle class lost up to 70% of their household income

Since 2009, the Greece incident has been repeated in different countries as in Venezuela and some countries in Africa. However, if you observe that common thread of these contemporary crises is the amount of increasing income inequality in favour of the high-income population. In certain countries, and due to human-made disasters and wars, this is causing continuous unsettlement and migration. This is precisely what is happening to the Syrians, the Iraqi's, the Afghanis. This same development is also happening in developed countries, particularly the United States, at an astounding rate.

Overnight chaos also affect the decline in mobility in advanced economies. This caused the most disadvantaged and the marginalized people. In developing countries as in the Arab world, the number of those living in relative poverty has risen significantly. The latest war in the Arab world increased the incidence of poverty diseases and malnutrition, especially in countries such as Syria and Yemen.

IMPORTANCE OF NEW NORMAL FORESIGHT TO OUR COMMUNITIES.

Foresight is a future intelligence-gathering that is systematic, participatory, and medium-to-long-term vision-building process aimed at present-day decisions and mobilising joint actions.

New normal socioeconomic foresight is based on sensitive research that uses quantitative and/or qualitative methods oriented towards the new normal at the junction of dream and reality aimed at shaping a more regenerative world. Qualitative socioeconomic foresight is useful for the elaboration of long-term visions having a broad socioeconomic scope in the new normal, such as medium- to long-term policy strategies.

Foresight and sustainable development goals.

The UN Sustainable Development Goals (SDGs) have listed 17 sustainable development goals (SDGs) to transform our world by 2030. There are eight direct socio-economic goals among these 17 goals which are: SDG1 and SDG2 which focus on zero poverty and zero hunger, then the SDG3 which focus on good health and wellbeing, followed by SDG4 which targets quality of education, then SDG5 which about gender equality, followed by SDG6 and SDG7 which are both about clean water, sanitation and affordable clean energy. Then last important goal among the top eight is the SDG8 which focus on decent work while maintaining economic growth. Graham (2019).

While these goals are right for the time, the new normal will see the rise of communities in leading on these measures. The goals move from nation state priorities to community priorities. Smaller, more agile entities must now move where nation-states find themselves in a political or diplomatic stand still. The power of the new normal will be in the rise of communities to take on the work where nation states may fail.

It has been a rallying cry by some companies and NGOs to point to the the reduction of extreme poverty rates around the world over the past 20 years. Despite this noble achievement, there are some challenging points that need to be raised. They include:

"Many believe they are changing the world when they may instead or also be protecting a system that is at the root of the problems they wish to solve."
Anand Giridharadas <u>Winners Take All</u> (2019)

The World Bank estimates that 40 million to 60 million people will fall into extreme poverty (under $1.90/day) in 2020, compared to 2019....Additionally, the percentage of people living on less than $3.20 a day could rise by 0.3 to 1.7 percentage points, to 23 percent or higher, an increase of some 40 million to

**150 million people. Finally, the percentage of people living on less than $5.50 a day could rise by 0.4 to 1.9 percentage points, to 42 percent or higher, an increase of around 70 million to 180 million people.
The World Bank, April 2020**

Poverty and how overnight chaos affects quality of life.

The latest socioeconomic crisis in Greece has led to incurred adverse effects on the health of the Greek population and has produced significant mental distress. In general, studies show a decrease in socioeconomic security has negative influences. For example, unstable socioeconomic conditions have made young adults delay marriage decisions and that reduced fertility rates and led to an increase also in divorce rates which caused further economic crises.

Creating better planning in the new normal.

We need to understand the gaps between the current and new normal status and rising demands due to fast changes coming up in the next 10 to 20 years. Through such foresight, we can reform the socioeconomic status towards a better-desired direction of tomorrow.

Many nation-states are planning their economies 70-100 years from now. These countries include Canada, Japan, Germany, France and Singapore. But the new normal is requiring communities to be just as forward thinking in their planning. These networked new normal communities will be able to share their ideas and methods with each other so that the distributed network model becomes stronger. This is not a substitute for nation-state planning, but a compliment to help make the entire system more agile and robust.

Therefore, an agreed national plan that brings in eclectic thought leaders, creatives, innovators, entrepreneurs, social activists, government representatives, academics and scientists can recharge a community into seeing itself in the future with a plan. Heading to a new normal requires understanding what and why before the how.

Re-engineering wealth.

Breaking or re-engineering any old tool of the old system requires an understanding of the rituals and the norms that create the structures, or the mechanisms of such system and its related dynamic effect. This means

You and the New Normal

we need to determine the drivers or the clusters that bring in such socioeconomic status. Thus, government policymakers can carry out national risk assessments that weigh the short to medium risks of the new normal foresighted socioeconomic disaster or disruption to human and economic welfare, to inform priorities for investment in preparedness and resilience.

One could imagine that the socioeconomic risk landscape is shaped by major external and internal trends of the known and the overnight chaos. These shocks in this landscape will increase if the resources are not appropriately optimized. Having stress from the growing demand for natural resources, due to climate change, would just one example of this. Understanding the hidden and the unexploited opportunities of our real wealth and assets would reduce the pressures of our socioeconomic crisis and would mitigate the effect of the rising frequencies of overnight chaos scenarios.

Any planning by a community needs to carry the probabilities of a positive and negative scenario by 2030 and beyond. Planning members need to determine the right resources needed to prevent overnight chaos scenarios. The actions should focus on eliminating old paradigm rules and concepts that simply don't apply in a modern world.

Thus, any structured or unstructured re-engineering of socioeconomic status would require two factors. First, a better understanding of the short- or long- term economic impacts of the abundant or scarce significant wealth of assets. Then, understand the natural influence of, for example, the environmental change, i.e. the influence of climate change, water scarcity, biodiversity loss, air pollution and the land-water-energy nexus.

Building a new normal in your community through 'Availability'.

In order to create a new normal model in your community, we need to link it to regional organizations and other communities who are already doing this level of work. Who is available to share their insights? Through focusing on availability, we can stimulate creative thinking and trigger reforms in and *around* institutional structures.

Learning to deal with current and long-term new normal socioeconomic challenges helps to shake-up the mindset and proposes designing new approaches based on the availability of what other organizations and communities have already tried and tested. This brings in better integration of different risks in analysis.

You'll find an eager number of organizations, experts and communities who have pursued new normal ideas, but found themselves alone in their pursuits. The new wrinkle in this process is that there is now a distributive network that you are creating to help implement those ideas that were once thought impractical.

Getting your people involved.

Building a new normal economy that is resilient in the face of overnight chaos or other challenges of the new world unfolding AND fully involves and engages humans, needs a total people involvement (TPI) technique that depends on human equality and partnerships. However, when we increase this level of involvement and move it towards total people engagement (TPE) through having the people design, plan and execute socioeconomic programs along with the decision-makers, then we can enhance the spirit of resilience.

The effective use of TPI and TPE could improve really the resilience behaviour of the communities and make them more capable of what to emphasize towards creating a better quality of life for all members of the community. This could help to shift the paradigm towards the holistic wealth that integrates with the process of factual decision making. For example, to encourage any for-profit social organizations that support the role of the civic involvement in the community economy, availability could be used to develop social control investments and funds that help to improve the economic development cycle again.

Establishing resilient new normal families.

In the new normal we can use numerous studies that focus on the role of family poverty to create more resilient family members. Studies, for example, have shown that some practices of poor parents help to promote socioeconomic resilience within the family. Even if a family goes through a divorce which eventually produces direct and indirect stress on their socioeconomic status, the availability of resilience-based social support from the community can reduce the negative impact of stress on the family members and could yield positive socioeconomic outcomes, such as producing entrepreneurs, or a productive family. The same would apply when the family would awake on the death of one of its family members. The community and its resilience mechanism would be the source of both recovery and maintaining a stable equilibrium which leads to both balance and harmony. Therefore, to establish resilient families, we must establish learning mechanisms that mitigate the effect of family disturbances and their influence on their socioeconomic status. Resilient family relations would help to reorganize the changing patterns of functioning thus to adapt to their new situation, where many social analysts would agree that it helps maintain or improve their quality of life and general wealth.

HOW DO YOU GET THE MIDDLE CLASS INVOLVED IN THIS PROCESS?

Transforming any community from being a receiver of what is planned, i.e. passive economy, as used to be during the peak of the industrial economy in 20th century; to an economy where people would interact with lots of creativity and collaborations is not an easy task.

Developed and emerging economy nations have maintained a history of the socioeconomically active middle class, where they exhibit their capacity to be resilient to any socioeconomic instability compared to people in developing and under-developed nations. The modern history, i.e. since the early 1900's, have shown that the main middle-class socioeconomic concerns are ar0und both socio-political changes and crisis in both the community and the country. However, the lessons that can be learned from developed countries is what happens when the middle class shrinks? This is usually a sign of increased income inequality, which leads to more for socioeconomic polarization. This is a product of old paradigm ideas and policy.

Engaging the middle class first means creating a new normal economic plan that provides a framework to build on. The values of collaboration, sharing, abundance and visionary leadership must be at the core of this conversation so that working class, middle class and poverty groups find common cause in creating a more robust quality of life. The new normal plan to build on must be first.

WHAT ABOUT AI?

With Artificial Intelligence (AI), we could eliminate many human challenges, but also, we could increase other problems such as alienation of specific human communities which could cause psychological problems. The latest rapid increase in the levels of depression and suicide are just one of the few signs of this new industrial development.

The 4th industrial revolution carries with it thus new types of poverty, that poverty in the ability to create effective decisions in comparison to AI Robots, which would limit our capacity to enjoy the well being we are striving for.

WHAT ABOUT YOUNG PEOPLE?

Cross-cultural studies have shown that youth participation in socioeconomic tasks, such as building on the new normal, is a great enabler for solving socioeconomic problems effectively. By enabling young adults into the community socioeconomic problem, we improve their self-esteem, enhance their moral development, increase their political activism and maintain their social relationships.

In education, if youth go through teaching approaches that are built around socioeconomic outcomes, their sustained engagement would be more guaranteed. With active experimentation or experiential learning, youth can start the curiosity journey of socioeconomic problem solving with positive mindsets. This positive mindset would make youth manage a time of constant change more effectively.

Throughout history, we see that there are groups of the society that may be marginalized in the time of prosperity, and they are the one who suffers most in times of crisis. Many women and youth who live below the poverty line would be the most affected during overnight chaos.

The need for youth had varied over the ages, when young people were a major force during wars, to periods when they became part of countries development advocates. As things have started to change and youth became the leaders of the economic change and the creators of ideas, besides the owners of wealth. However, this did not reduce poverty among young people.

WHAT ARE INSPIRATION LABS?

Inspiration labs are "boots on the ground" opportunities to test the challenges and opportunities of a new socioeconomic way forward. It is a way to explore the hidden opportunities inside the socioeconomic problem relevant to issues as in health care, education, community needs, youth and women empowerment and get the results in real time. Below is the data from one of those labs.

CASE ANALYSIS AND FINDINGS.

This chapter was created to identify the relationship between socioeconomic issues and new normal foresight. The results show how we can transform socioeconomic issues proactively, through models that would either mitigate a crisis, or prevent its occurrence, or learn from its outcome once it is over. The idea here is to ensure that we create sustainable models for the new normal foresighted crisis.

Ten cases were selected. Then the cases were evaluated as to how the inspiration labs helped to mitigate their actual coming risks, or how they create learnings, or lessons and opportunities for improvement after the actual socioeconomic crisis.

In Table (4) shows how the specific selection of inspiration labs tackled socioeconomic issues are linked to specific probability and hazard that have 3 levels (high, medium and low). Based on the type of the calculated risk which comes from the results of Probability x Hazard, we can visualize the new normal foresight of the socioeconomic issue and prevent crisis (PC), or mitigate crisis (MC).

Table (4) illustrates the probability of the influence of the socioeconomic issues and their hazards if not tackled in the right way and right time. The table define what type of crisis new normal foresight the Inspiration Labs (ILs) are helping us to achieve. i.e. what the specific type of lab is leading to? The framework is mainly only two identified crisis management tracks, either prevention of crisis (PC) or mitigation of crisis (MC).

Table (4) Links between socioeconomic Issues and Possible type of Crisis Management

Socio-Economic Issue tackled by ILs	Probability of its Influence	Hazard if not tackled	Type of Foresight
1. Raising Rural Communities & People in the Slums access to Education or basic life necessity schools	**High** (i.e. more people can access education, and more types of education (formal and informal) found the fewer poverty variables would exist.	**High** (more drastic conditions of absolute poverty and people working in very poor conditions)	(PC)
2. Developing the Quality of Life of the Elderly (eliminating elderly homes and enhancing Geriatric Care Homes)	**High** (Longer quality of life and more integration between geriatric healthcare services and the family)	**High** -Majority of the population would not be fully functional and would need more care. -We would have lonelier elderly people -Bigger gaps between generations as youth won't meet their grandparents	(PC)

Socio-Economic Issue tackled by ILs	Probability of its Influence	Hazard if not tackled	Type of Foresight
3. Improving the contribution and the competitiveness of the retired by redirecting the investment of the pension funds	**High** - Maintain both the knowledge and the value of the retired	**Medium** We would witness more depression cases	(PC)
4. Adopting multi-purpose public buildings for schools, healthcare centre and events gathering area	**Medium** - more utilization of building - better for the environment	**Medium** -Minimising the probability of more isolations	(MC)
5. Creating a Model of Poverty Elimination with minimal resources	**High** -Reduce the burden of targeting zero absolute poverty in developing or under-developed countries	**High** Minimise the speed of volatility of the marginalised communities	(MC)

Socio-Economic Issue tackled by ILs	Probability of its Influence	Hazard if not tackled	Type of Foresight
6. Establishing Early Students micro start companies	**High** Entry to entrepreneurship would enhance the levels of innovation and creativity of youth	**Low** Spirit of entrepreneurs would stay something that is measured at later stages of life	(PC)
7. Enhancing the impact of 'woman development', not only 'women-empower.'	**High** -The progress of women who will help the progress of society	**Medium** Society would still see women as the weak part of the community	(PC)
8. Eliminating Water loss	**High** -The capacity to discover early the water leakages with minimal resources would ensure effective maintenance of clean water	**Medium** Water would be a scarcer resource.	(MC)

Socio-Economic Issue tackled by ILs	Probability of its Influence	Hazard if not tackled	Type of Foresight
9. Early detection of Non-Communicable Diseases (NCD's), i.e. Diabetes, Blood Pressure, Cholesterol and Obesity in the population without detailed examination	**High** -More quality of life -Fewer patients with complicated cases	**High** -More early death -More people with limited productivity	(MC)
10. Establishing Inspection' that minimize the rates of poisonous food with minimal intervention	**High** -More consumers trust and fewer incidents of food diseases	**Medium** Poisoning may increase due to lack of human immunity Resorting to supplements that have complications	(MC)

Dr Mohamed Buheji & Futurist Chet W. Sisk

Patterns of resilience during new normal socioeconomic crises.

Since 2008, the world has been exposed to the socioeconomic crisis that requires solidarity and psychological analysis. Developing socioeconomic resilience frameworks by testing it through problem-solving labs helps to blend social, economic and cultural practices and the new normal foresighted challenges, and this could minimize the suffering from coming harmful or hazardous situations. Buheji (2018b).

This chapter focused on sustainable patterns of both coping and adaption. Through the synthesis of the ten cases, we can see that inspiration labs and similar live labs could help to the foresight and work on many crises early and thus either prevent it or mitigate it once occurred. This resilience to socioeconomic crises needs to be supported by effective re-designing both the community cultural and social activities. Buheji (2018b).

Benefits from the labs.

The most relevant aspect of this study is that it shows that the new normal socioeconomic challenges can be predicted. Practical field labs, as the international inspiration economy project (IIEP), inspiration labs (ILs), helps to develop the community capacity to deal with the new normal socioeconomic crisis. Therefore, the chapter calls indirectly for more such initiative as IIEP (ILs) that bring about realized positive socioeconomic changes with clear, experienced outcomes. The researchers thus recommend that the reference models differentiate by initiatives and approaches as IIEP and ILs should be studied and developed further to help crack out any new normal human problems with minimal resources.

The approach that ILs projects bring could be generically carried by institutions, non-government associations, societies, private sectors, and then publicized through social media, books publications and scientific journals, so that many communities could avoid such new normal problems. The worked upon the new normal foresighted socioeconomic problems could also be a source of development of communities and organizations and could help prevent the following types of poverty, complicated youth migration, slowness of quality of life and similar challenges.

CHAPTER 14

WHY CURIOSITY IS SO IMPORTANT IN THE NEW NORMAL?

Curiosity always has been linked to exploration for a hidden passion, query or inspiration insights. Today, it is even more linked to a life-journey style. It paves the road on the path to possibilities.

Curiosity is measured by the amount of IPI (Imaginal Processes Inventory). IPI reflects the level of curiosity that we usually see in questioning, investigating, trying, testing and then possibly learning, but not necessarily about getting the right answers. Therefore, to reach such levels of curiosity, we need to have equity, diversity and inclusion about what we are curious about.

To sustain the mindset for curiosity, we should start our enquiry with 'why'. Curiosity is associated with direct utility from information and is defined formally by using the concept of entropy.

THE CURIOSITY ECONOMY.

We live in a world controlled by a capital-driven economy where the economy wants to thrive, whether there is real growth or not. This capital economy does not address the human needs to thrive, whether there is growth or not. To address this thrive, we need to be curious.

Curiosity is of high demand in today's disruptive and fragmented world. Therefore, this research looks at how curiosity-driven practices influence any community innovation and its economy. More precisely, how curiosity-driven explorations can lead communities to a new normal we'll like.

Freeing the mindset in the new normal from fixed perspectives would help us to deal with opportunities and challenges more effectively. The ability to overcome our assumptions also would help to be capable of pre-determining the expected solutions.

If a community has the 'curiosity to discover' it would have the capacity to overcome the 'complacency trap'.

Figure (13) Overcoming the Curiosity Economy Traps

Designed curiosity.

Curiosity about new normal extraordinary neurological connections helps us to use our imagination to develop our new potential outcomes. Curiosity about the new normal triggers our imagination to reflect on some possible scenarios that come from novel approaches, data and a sense of imagination.

When we are curious, we generate and play with new normal ideas. Our foresight capability in the new normal can help us to try new things and acquire new experiences constantly. Our new normal drive makes us always eager to know different people, and connect to patterns that other people cannot see. New normal foresight makes us more persistent; because we want to get to achieve something unforeseen and with no evidence. This type of new normal curiosity drives us to a higher sense of mission, purpose and meaning.

> Figure (14) shows the different five main capitals of wealth that are used for new normal re-invention and future foresight.

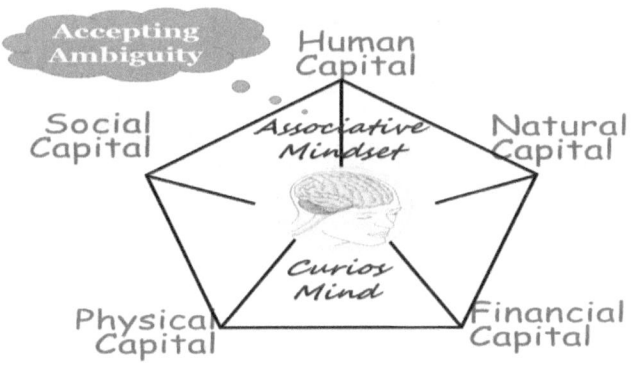

RE-OPENING THE CURIOUS MIND.

One of the reasons we want to make sure we drive home the point of curiosity is because we recognize how important it will be to the new education of children. Children have a natural curious nature that needs to be supported and developed. Even more, leaders and executives can be retrained to capture their natural curiosity once embraced in their youth. This model shows how the curious mind works

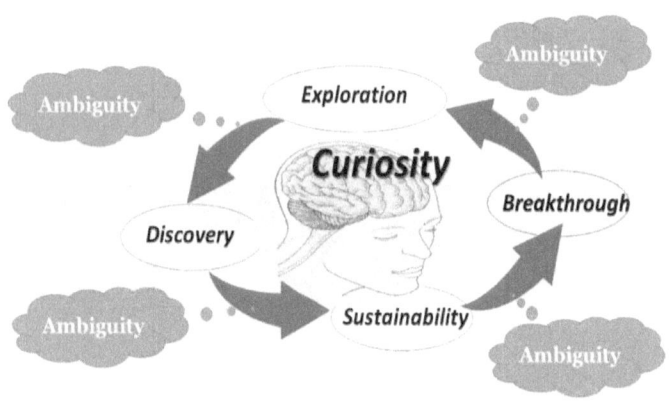

Figure (15) Cycle of Curiosity in Inspiration Economy Projects

IMPORTANCE OF REALIZING THE CURIOSITY FOR NEW NORMAL FORESIGHT.

Curiosity is a trait that can be seen in almost all people before they reach the age of seven years. However, this curiosity dissolves once most of us become moulded by the educational system, our parents, or the community surrounding us. Most of us, by the time we finish 6th

grade would have gone through a setup or a mindset and a surrounding that undermine what we used to believe in or imagine. Thus, over time we start to believe that what we dream in, or what we used to imagine not realistic. Our minds then become so trapped with many questions. We want to explore and learn about the world we are living in or the challenges of the new normal, but the fear of releasing these types of questions increases over time.

In the meanwhile, today, we still live in a world mostly being more concerned about getting the right answer than asking disruptive questions. Therefore, the new normal will need us to champion disruptive questions. The foresight of the new normal will not depend on categorizing people based on standardized tests or classifying them based on grades or achievements, but rather on practices of explorations or attempts of differentiated contributions.

The world started to appreciate not only our level of curiosity, but also what we are curious about. Specifying what fascinates us became a consistent question in any job interview. Following our passion is becoming highly appreciated by leading employers of today. We are actually expected to consistently nurture our curiosity with new discoveries and create new solutions for complex problems around us.

The new normal will be fertile ground for the "mavericks" --- those that do not like to be systematized, or told to obey orders without much questioning. What worked for a mass-production industry, does not work the fourth or fifth and future sixth industrial revolution, which depends on workers creativity and welling to be much more than being proactive. Curiosity is used more today in managing the dynamic changes, and this raises our imagination.

The process of new normal foresight is similar to curiosity pathways. In both, curiosity and new normal foresight, we go through first observing "humbly" the possible interactions and then exploring and closely defining the scope. Then, learning from the interactions happens which improve the capacity to visualize again.

CURIOSITY RAISES OUR ABILITY TO CREATE SOLUTIONS.

Curiosity' drives the problem solver to areas never thought of. Curiosity makes us dare to experiment in totally different communities and in different times, Buheji (2019a). Therefore, with curiosity queries, we can provide foresight about what makes the socioeconomic problem different—for example, being curious about why do most the youth would continue to migrate in the new normal. Thus, foresight for the potential social change makes us more curios in finding the right problem, besides being curios of finding the right solution. UNDP (2019), Eggers and Macmillan (2013).

Through this alternating finding the right problem or solution, we can build the excitement or the challenge towards the new normal foresight. At certain stages, we need to test the outcome in order to deliver a sustainable outcome, as shown in Figure (16).

Figure (16) Alternating between Curiosity for Right Problem and Right Solution

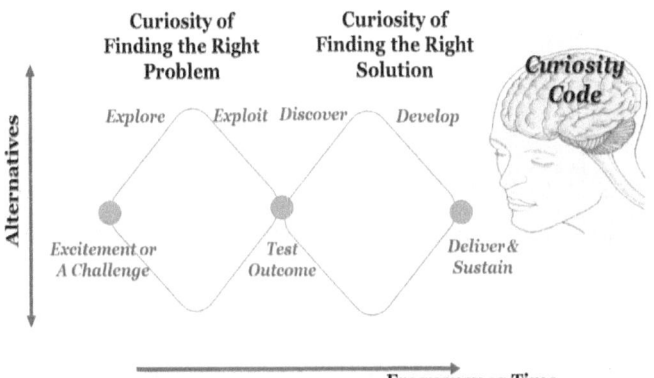

When organisations and societies become curious, they create meaningful, unique and diverse opportunities. With curiosity, we can improve lateral thinking to deal with complex problems which lead to communication improvement. The availability of a suitable kind of

curiosity enhances the understanding of customers' needs through completing the ideas and building synergy.

Curiosity Types.

There are many different kinds and types of curiosity. This is due to curiosity can come in a whole variety of qualities and wavelengths, flavours and intensities. The different types of curiosity are the key to understanding people's personalities and motivations. These different types of curiosity create the capacity of new normal foresight.

Curiosity can trigger new normal foresight by different means. The different curiosity types would motivate different kinds of questions and this help to build a sense of purpose. The following curiosity types are selected based on a previous work of the researcher, Buheji (2019a). A framework for using curiosity and its economic constructs for more accurate new normal foresight is later proposed.

'Social' Curiosity.

Social curiosity is about being involved with social issues and being able to connect effectively during a social gathering or events. Social curiosity helps to discover opportunities ingrained within the community. Many of the social curiosity was the cause of many changes in history. When we are inspired by a greater purpose in life, or new ideas, or projects that come as a result of this social curiosity, we actually help our thoughts and mind to transcend beyond limitations.

'Emotional' Curiosity.

Curiosity can be described as positive emotions towards acquiring knowledge. Discovering new information may also be rewarding because it can help reduce undesirable states of uncertainty. Besides, curiosity is seen as a pleasurable experience that comes with exploratory emotional behaviors. Emotional curiosity can be seen with thinkers or leaders that become obsessed with creating a definitive change.

'Problem Ambiguity' Curiosity.

Living with ambiguity in the new normal means, we will have an open mindset that manages the challenge of the new hypotheses while attempting to reduce the blind spots. Through ambiguity of any problem faced in life, we can build curiosity and thus avoid any negative thoughts. Neuroscience now shows that not knowing the answers actually inspires the human mindset as not knowing the clear path, or even the outcome of a socioeconomic problem when we start, raises our curiosity for exploring and learning. This would help us to come up with creative solutions and pursue the outcome that we could not clearly visualize.

'Lifelong Learning' Curiosity.

The advancement of neuroscience is making us appreciate the different constructs of curiosity. Now we know more than ever that curiosity can come from the state or the condition we are in, i.e. comes from extrinsic conditions, or due to our intrinsic traits, i.e. we can determine when curiosity occurs.

'Exploration then Exploitation' Curiosity.

Exploration means searching for unknown or unfamiliar areas. You usually need a research technique or field study with a spirit of enquiry to achieve successful exploration. Exploitation, on the other hand, need intentional actions that target to benefit from resources around us. This probabilistic element causes them to occasionally explore other possibilities, leading them to better overall choices.

'RESILIENCE' CURIOSITY.

Resilience curiosity flourishes when we start the reasoning stage of what we experience. With active experimentation, or experiential learning, we start the curiosity journey of socioeconomic problem solving and a positive mindset. Once inspired, people, organizations and societies become more resilient and tend to be more curious to explore the inner strengths and focus on outcomes while optimizing resources. This applies again to organizations and societies, even more individuals.

'INNOVATION' CURIOSITY.

'Innovation' driven curiosity can come from different dimensions. Excessive anxiety impedes human lust for new discovery. Curiosity brings in multidimensional creativity model that can be integrated into a single model. With 'innovation' curiosity, we recognize and seek-out new information and experiences. In creativity and innovation, the curiosity on the problem triggers the absorption dimension as a tendency to be fully engaged in activities.

'Productivity Focused' Curiosity.

If we study the productivity of most organizations, it is usually linked to its capability to mobilize people and resources towards specific goals. Rarely we would see that the productivity of the organization focused firstly and mostly on capacity. When organizations focus on developing its capacity, instead of making its goals, its capacity for discovering new opportunities will link curiosity with life-fulfilment goals, and this would influence the organisation current and new normal performance.

'Intellectually Stimulated' Curiosity.

Intellectually stimulated curiosity has been defined as instructors' ability to challenge students and promote intellectual growth. Possible approaches in the new normal can be through using an interactive teaching style. This style engages and challenge the students (e.g., making sure students know the material well and pushing students to do their best), or encouraging independent thoughts. For example, in the new normal we can help students to think critically and come to their own conclusions.

'Empathetic' Curiosity

Empathetic Curiosity can only be appreciated by experimenting and action, not by perception. When we actively experiment links of cognition within a specific environment, we enhance our perspectives of the problem investigated, and then our field contribution can be differentiated. In order to reach empathetic curiosity, we need to move our curiosity and exploration from focusing on money-making to meaning-making.

Case Study

Brief on the Inspiration Economy.

The inspiration economy focuses on tackling issues relevant to poverty, youth and women empowerment, unemployment, migration, community productivity, family stability, community diseases, inter-generation challenges, etc. Buheji (2018b).

Since September 2015, the IIEP has tackled more than 300 types of projects in more than 13 countries. For the scope of this study, only seven different cases were selected to test the implication of the related curiosity type on the new normal foresight of the socio-economic. Buheji (2018a).

Defining the Normal Foresight Based on the Type of Curiosity.

Table (5) set up the relation between socio-economic issues, the type of curiosity and the implications of the foresighted new normal.

Table (5) Illustrate the New Normal Foresight as Per the Curiosity Type

Socio-economic Issues	Type of Curiosity	Implications on the Foresighted New normal
1-Reducing Levels of Chronic Anxiety among the different generations	Social/ Emotional/ Lifelong Learning/ Resilience/Productivity Focused/Empathetic	Collective Social Applications that enhance the accuracy of the type of anxiety and calibrate the beneficiary
2-Intergeneration Gap	Social / Emotional / Lifelong Learning Exploration then Exploitation/ Resilience/ Intellectually Stimulated Empathetic	Social integration programs that bring more inter-generations together to solve a common good issue, i.e. socio-economic challenge and be engaged with it

3-Women are the highest Graduates and Highest Unemployed	Social/Emotional/ Problem Ambiguity/ Resilience/Innovation Productivity Focused Intellectually Stimulated Empathetic	Enhance early women business and social entrepreneurship programs
4- Non-Communicable Diseases (NCD's) (Diabetes, Blood Pressure, Cholesterol and Obesity)	Social/Emotional/ Problem Ambiguity/ Lifelong Learning/ Exploration then Exploitation/ Resilience/Innovation Productivity Focused Empathetic	Enhance the capacity for early discovery of NCDs and Keep Calibrating LifeStyle as per demographics
5-Low Profit Margin of the Poor	Social/Emotional/ Problem Ambiguity/ Lifelong Learning/ Exploration then Exploitation/ Resilience/Innovation/ Productivity Focused/ Intellectually Stimulated/Empathetic	Social-for-Profit Projects to end the Multidimensional Deprivation & Multidimensional Poverty through creating Mediators that enhance the Profit Margin of the Poor
6-Limitation of Water in Dry Season causing Hunger & Draught	Social/ Emotional/ Problem Ambiguity/ Resilience/Innovation Productivity Focused/ Empathetic	Improve well and dams management and re-evaluate the type of products and farming

7-Open defecation Problem	Social/Problem Ambiguity/ Lifelong Learning/ Exploration then Exploitation Resilience/Empathetic	Enhance the capability of the people to deal with their waste and defecation in safe areas and ways. Then, enhance training for linking these safe areas with toilets. Training the trainer for children in schools and then put free Well maintained public toilets in the villages and communities.

FRAMEWORK FOR ENHANCING CURIOSITY OF NEW NORMAL.

Being economically curios means we can build a mindset that can visualize possibilities in a new environment with new tools. Exactly what's needed in the new normal. Figure (17) represents the framework proposed for further studies. The framework comes as a result of the synthesis of the literature, and the case study presented. More test of the proposed framework in other fields than the socio-economic problem solving is highly recommended.

Figure (17) Framework for Enhancing Curiosity of New Normal Foresight

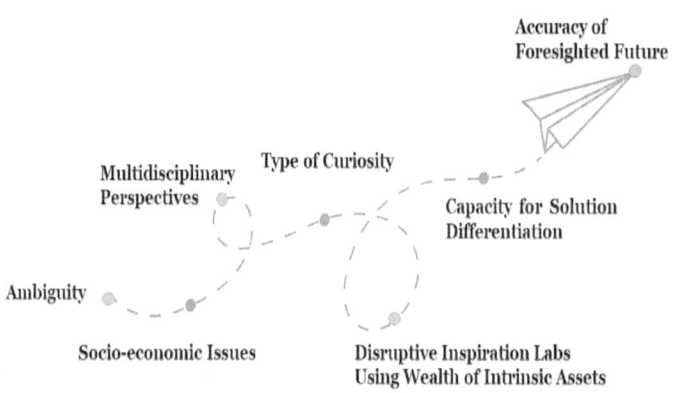

Understanding the power of curiosity will be one of the defining elements of the new normal, as we move from cogs in a machine to thinking, imagining, visualizing beings.

CHAPTER 15

A PHILOSOPHICAL TAKE ON THIS MOMENT IN TIME.

Futurist Chet W. Sisk

Make no mistake. The global shift in how things work is happening right now. It is a verifiable, evidence-based, measurable phenomena happening in real-time, and it requires an entirely new set of approaches to successfully navigate this transition period. COVID-19, as well as climate change, are addendum onto an agenda of transformation. Any doubt about this is simply a denial of the facts and the trends as they exist. I'm not the only one who has arrived at this conclusion:

"Asia will enjoy the global power status it last had in the Middle Ages, while the 350-year rise of the West will be largely reversed. Global leadership may be shared, and the world is likely to be democratizing. It all depends on how

events develop over the next decade. We are at a critical juncture in human history, which could lead to widely contrasting futures."

Global Trends 2030, the National Intelligence Council 2012

The National Intelligence Council is a gathering of intelligence organizations in the US and around the world, assessing the data and making projections. While I see things a bit differently than their view (Asia will not rise using the same tactics of the past 350 years), we agree on the core – we are in the heart of transformation.

What we do right now, over the next few years, will affect the world for centuries to come. The key part of the paragraph you just read is that we are standing at decision time for humanity. Using the tools that we outlined earlier, we must choose what kind of future we wish for us, our children and our children's children. Our job as both future foresight experts and futurist is to make sure you know the path to a more empowered and abundant world we've been talking about in legend and lore for aeons has been meticulously laid out by some hardworking individuals and organizations. Sometimes it's hard to see a hurricane when you're standing in the eye of the storm. However, that's exactly where we are.

You'll notice that we didn't spend a lot of time talking about the new technologies as part of this transition period. That was not an accident. The technology is important, and it will help redefine our future. However, the key to the future is not technology, but what we believe the technology should be used for. Technology should be an extension of our greatest dreams and aspirations. It doesn't lead. It simply is the tool we use when we lead.

From the experience of the authors as experts who have done different projects in relevant to technology, it can be said with great confidence that technology without thoughtful, contemplative, higher thought people behind it, creates a ticking time bomb. We still hear people talk about how to weaponize the technology or how it can spy on us and track us and know our every move and predict our behaviour. Those people weren't talking about how we can end poverty or hunger. The environment in the old paradigm we are now leaving with the new technologies was consistent with giving the keys of a Lamborghini Aventador to a 5-year-old. The child may have a *concept* of how the vehicle works, but the child doesn't have the size, maturity or capacity to be behind the wheel of that car. Even in the face of futurist/inventor Ray Kurzweil's wildest fantasies, our left-brain technology development won't save us. Our human approach to the road ahead will save us. The technology will simply be an outgrowth of our new collective consciousness that marries both left and right brain with our intuition. Putting the development of our technology ahead of the development of us is the same as information without understanding, intellect without discipline, opinion without responsibility.

I believe that my writing serves as a bearer of good tidings. It is not beyond us to create a more efficient and effective planet while saving the planet from our excesses. Yes, many people can see this new world ahead, but we now are at the point where we need doers.... champions of the idea that we can do better than this. They must be bold enough to lead from a new position.

The best way to make things work in the new world is if we double down on becoming new people. The reverse is also true. The world will become increasingly difficult for those seeking to hold onto the ideas of old. The climate change challenge itself will do some weeding of its own, asking us to make hard decisions of how we are going to manage life on earth. It appears that our transition and the challenges ahead are happening at the right time. As the opportunity for us to become our greatest selves appear, the challenges that need our greatest selves also emerge.

Yes, the world is transforming, but that doesn't mean there aren't those who have a vested interest in maintaining the old approaches. Trying to apply old paradigm concepts in a new world will lead to frustration and chaos. Only those who become something bigger and more profound will fully experience the power of this time. The same principle applies in this new chapter of humanity. If one seeks to live fully in a transformed world and reap the benefits and rewards of it, then they too must be transformed. It is hoped that the approaches outlined in my writings can lead to systemic as well as personal renewal.

This position of transformation will be disturbing to some because it may appear politically inconvenient. Many will seek to paint this writing and its contents as simple political conjecture with an agenda. Of course, we would disagree. This is a conversation that is past due, but it needs infrastructure, so we know what to do. It is an assessment based on the data in front of us. That data keeps asking us to as one central question: what's happening? Once we were able to get that initial question answered, we followed up with two others – what does all of this mean and what are we to do now? No particular political party, politician or organization that has an exclusive monopoly on answers to those questions.

My 'Family Council' experience with my family of origin.

One of the most important developments of my life was something that my parents called Family Council. It was a vehicle created by my parents to help the household work better. My parents --- James and Naomi Sisk, along with my four siblings ---- Denise, Jerome, Wade and Rodney, lived inside a virtual new paradigm laboratory in my formative years. In Family Council, our family would meet several times a month to help guide the family's direction. These meetings gave all the family members an equal platform to air their grievances, share their successes, vote on vacations, receive an assignment of chores and were allocated allowances. This was my exposure to lateral, personally empowering organizations. It has been a part of my thinking all of my life. This was unprecedented in the 1960s and 1970s for households in general and exceptional among African American families during that time. I also had the chance to witness my mother and father team up as partners as they ran the household and created entrepreneurial businesses together. This first exposed me to the rule of interdependence, where their collective challenges and collective successes were shared. Also, their commitment to entrepreneurship exposed me to the power of following my dreams through the sweat of my brow and the insight of my thinking. This helped me to fully embrace the idea that we, all of us, have the tools of creating something special in the world if the resources that create success-producing environments are shared through equity.

Context

One of the most important parts of this transition period is to make sure we apply context to what we're seeing. History has plenty of those. Think about the life of the woman who lived during the European dark ages. Barely enough for her to eat, let alone for her family, or the African who died of disease and starvation as he lay in chains on the slave ship, or the Jewish elder who was drawn and quartered because he would not submit to the pressure of the church and renounce his religion during the Inquisition, or the young boy who was beheaded because it was deemed so by the Khmer Rouge, or the woman who was tortured and impaled during the rule of Vlad the Impaler. To say the world has been an unsavoury place in the past is not giving justice to what we just walked through.

We are literally walking out of one version of earth and into another version. This transition period bears a striking resemblance to so many predictions in sacred scriptures from faiths around the world. We would leave lives of pain and injustice and into a new world of opportunity, empowerment and abundance. What we're discovering from the data is that door has been in front of us for some time. Only now are we realizing that it's time to walk through. Yes, we may wax poetically about what is to come, but those who ignore the data shared in this book do it wilfully in order to stay with the narrative they've been told all of their lives. They certainly can choose to ignore the obvious, but that doesn't make it go away.

THE SEASON OF THE NEW NORMAL.

I have a deep appreciation for a particular cycle -- the changing seasons. Summer, fall, winter and spring all have aspects in our life; all are truly loved and enjoyed. The heat of summer often provided the foundation for grand thunderstorms, the light shows they produced, those epic all-day journeys on our bikes and then come the cool of the evening. The fall provided us with the vibrant colours from the changing of the leaves on the trees and the fresh harvest from the neighbourhood garden we all benefited from. While the winters were cold, the beauty of the snow provided a rich background during the holiday season and ample ammunition for a good snowball fight. Then there was spring. Spring always brought the promise of renewal, growth and possibilities. Even when we were in the middle of winter, the promise of Spring gave us the energy to anticipate warming temperatures, the greening of our environment and a good feeling from neighbours who were ready to emerge from "cabin fever."

Sometimes during the budding of the plants, when the Honeysuckle trees and Lilac bushes were in first bloom, there would be a residual late winter storm that would come out of the blue with colder temperatures and snow to cover those fragrant blossoms. Some branches would break and the early growth out of the neighbourhood garden would be blanketed. Even with this storm, we still knew Winter was on its way out. We still knew in our hearts and minds that this was spring. Everything around us told us this was true. The late-winter storm is just a parting shot and a reminder of what we just left.

That is how we see what is happening in the world right now. Spring is here. There are fundamental transformations going on in how humanity operates all around the world. As with early spring, there will be a few late Winter storms of negative and unfortunate circumstances that will happen in the world during this period. The COVID-19 epidemic is a perfect case in point. The storm may break some branches of new movements and even cover some of the new blossoms of hope in our

You and the New Normal

societies, but don't be fooled by appearances..... *it is still spring.* We are simply experiencing another form of the cycle pattern found throughout the known universe. As it is early spring, we also must live in two realities. We know the seasons have changed and it is time to plant, but we must be diligent tillers of the land --- caring for the garden so that the new growth has the best chance possible of producing a bumper crop under our stewardship.

The final desire of this book is to inspire you, but it is also a call to action. Hopefully, the examples you've read of these "farmers" going to work, or have inspired you to also step into your responsibility of planting and tilling. We will see a great harvest, but only if we are diligent in our work now, while the ground is thawing and fresh with nutrients. There may be some things you've seen in the world you know we can do better. In the old paradigm, you waited for someone else to provide an answer on how to solve the problem. In the early spring of the new paradigm, the new normal, you need to step into that leadership and do the work and collaborate with others on how to do it more effectively. It's no longer just you trying to make something work. You are now going downriver, with the flow of the water and with the wind at your back.

We are now living in time for the innovators, the visionaries and the seers. Remaking how the world works is not easy, but we have everything we need to get the job done.

It is often stated that fortune favors the bold.
I would submit, it also befriends the visionaries.

CHAPTER 16

WHAT COMES AFTER SURVIVING HUMAN COMPLEX CHALLENGES?

Dr. Mohammed Buheji

The challenges and opportunities of behaviour change are found more than ever today from the complexity of life challenges. The transformation towards the new normal would create problems to be solved; these problems are full of complex formulas and complexities that are full of ambiguity that is yet to be discovered. But we must believe that it can be discovered.

The impact of new normal on socioeconomic challenges specifically requires behavioural interventions that tackle complex situations. Behavioural interventions will be only one tool towards creating social change. However, it is a very effective tool if it is gained by 'learning by doing', or by 'learning by exploring', this what builds the multi-disciplinary solutions that lead to the 'new normal'.

The new normal would keep people, organizations, communities and governments busy with the spillovers of the posit-COVID-19 pandemic, however also this would come with both top-down influence, i.e. from government and big companies, and bottom-up led by the civil society influence. Figure (18) the new normal influencers and where we focused on this book. We believe that those who are going to do a major influence in the new normal are those who realized these constructs and optimized them.

Figure (18) The New Normal Constructs and the focus of this Book

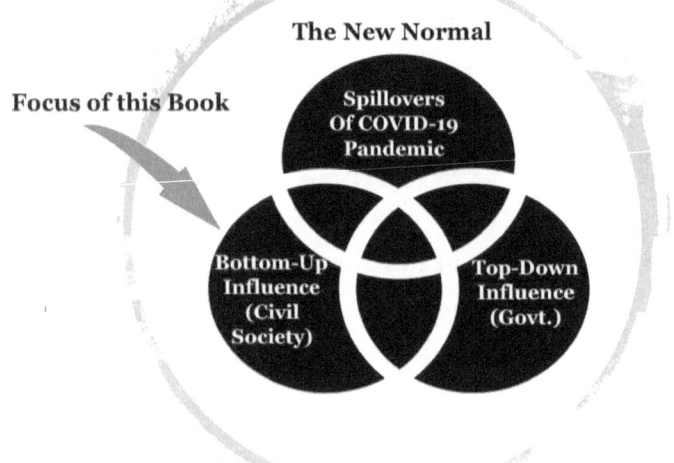

You and the New Normal

The new normal would carry lots of social system problems which are ill-formulated and confusing, but this confusion is what going to cause the mind to be flouting and searching for opportunities that can't be seen otherwise. This challenge would bring more engagement from people of conflicting values. This brings in better "solutions" that harness the community's capacities.

THINK ABOUT THE POSSIBILITIES WAITING FOR YOU.

There is a very simple equation that has led to this book….change, or suffer the consequences. This book was not written to scare you with that very real assessment of the immediate future, but to lead you to think about the possibilities that await you, once you decide that the potential reward of the future is worth taking the extra step, doing that extra thing, summoning that extra courage.

There is, indeed, a 'new normal' emerging in the world. That new normal bares the gifts of merging modern technology with indigenous wisdom with an optimistic vision for the future.

The research brought forward in this book is meant to be simple, however, it should remind us that the data is on the side of transformation, that opportunity awaits those who can see through the muck and mire and that this is a fantastic time to be innovative, intuitive and adaptive.

We hope that our work validated those of you who knew there was another side to the world presented to us, but didn't know how to find it. Indeed, that world has always been at our fingertips, but was obscured by things and images that were not our own. See you in the field.

BRIEF ABOUT THE AUTHORS

Mohamed Buheji

Dr Buheji is the founders of the International Inspirational Economy Project and Institutes. He is considered a leading expert in the areas of Inspiration, Excellence, Knowledge, Innovation, Inspiration, Change Management and enhancement of Competitiveness for over 25 years. He is a retired professor from the University of Bahrain. Besides being a Future Foresighter.

Dr Buheji is also the Founder of the International Journal of Inspiration & Resilience Economy and International Journal of Youth Economy. He has published since 2008 more than 70 peer-reviewed journal and conference papers and 17 books in the subject of the power of thinking, lifelong learning, quality of life, inspiration and competitiveness. Also, he has five books in English about Knowledge-Economy, Inspiration Economy, Inspiring Government and Inspiration Engineering, Resilience Economy and Youth Economy.

He is passionate about transferring his + 500 consultancy projects experience for more than 300 organisations from all over the world, to both education and research.

Besides, Dr Buheji serves in the editorial board of 5 internationally peer-reviewed journals. He is a member of many scientific communities, journals, academic review boards. Lately, he is a winner of many awards including the latest CEEMAN best researcher award for 2017, besides being a Fellow of World Academy of Productivity Science.

CHET W. SISK

Chet is a futurist, entrepreneur, author, consultant, speaker and workshop developer. He helps organizations see through the crisis of the moment in order to create plans for the immediate future. Chet's focus is on developing organizations and individuals that can thrive and succeed in a time of massive change due to climate change, technological evolution and social disruption. He is a member of the International Institute of Inspiration Economy, an organization designed to inspire economies that engage the public and inspire innovation, empowerment and possibilities.

Chet has spoken on these subjects at the United Nations in 2015. He has presented in almost 30 different countries, including South Africa, Albania, Malaysia, Nigeria, The Netherlands and Bahrain. He has taught at universities around the world including The University of Kwazulu-Natal in Durban and Pietermaritzburg, South Africa as well as at Al-Akhawayn University in Ifrane, Morocco.

Chet is also the founder and principal of Universal Basic Resources, a consulting group focused on helping develop climate strong companies and organizations so that they enjoy success in a climate affected world. He is a consultant for the city of Denver, Colorado, USA as it seeks ways to develop fiduciary, logistical and social response to potential climate change affects.

Chet written several books on personal and corporate change, he was born and raised in Waterloo, Iowa in the United States.

www.ingramcontent.com/pod-product-compliance
Lightning Source LLC
Chambersburg PA
CBHW020635220526
45464CB00001B/151